Ben Stacy Jerrik (Ed.)

Sawbridge

Ben Stacy Jerrik (Ed.)

Sawbridge

Warwickshire, West Midlands (region), England, Daventry, Rugby, Warwickshire, Grand Union Canal

Part Press

Imprint

Publisher:
Part Press is a trademark of
International Book Market Service Ltd., 17 Rue Meldrum, Beau Bassin, 1713-01 Mauritius
Email: info@bookmarketservice.com
Website: www.bookmarketservice.com

Published in 2013

Printed in: U.S.A., U.K., Germany. This book was not produced in Mauritius.

ISBN: 978-620-1-58093-0

Contents

Articles

References

Sewickley_Heights,_Pennsylvania

Sewickley Heights is a borough in Allegheny County, Pennsylvania, United States. The population was 810 at the 2010 census.[] Sewickley Heights is one of the wealthiest municipalities in Pennsylvania.

Geography

Sewickley Heights is located at (40.561091, -80.155541)[] .

According to the United States Census Bureau, the borough has a total area of 7.3 square miles (km^2), all of it land.

Demographics

As of the census[] of 2000, there were 981 people, 336 households, and 273 families residing in the borough. The population density was 133.9 people per square mile (51.7/km²). There were 355 housing units at an average density of 48.5 per square mile (18.7/km²). The racial makeup of the borough was 97.35% White, 0.92% African American, 0.61% Asian, 0.20% from other races, and 0.92% from two or more races. Hispanic or Latino of any race were 0.31% of the population.

There were 336 households out of which 29.5% had children under the age of 18 living with them, 75.0% were married couples living together, 4.8% had a female householder with no husband present, and 18.5% were non-families. 16.1% of all households were made up of individuals and 6.8% had someone living alone who was 65 years of age or older. The average household size was 2.56 and the average family size was 2.87.

In the borough the population was spread out with 20.6% under the age of 18, 3.5% from 18 to 24, 17.8% from 25 to 44, 30.1% from 45 to 64, and 28.0% who were 65 years of age or older. The median age was 50 years. For every 100 females there were 74.6 males. For every 100 females age 18 and over, there were 74.3 males.

The median income for a household in the borough was $115,672, and the median income for a family was $158,756. Males had a median income of $89,473 versus $40,417 for females. The per capita income for the borough was $79,541 placing it at #99 on the list of highest-income places in the United States. About 5.2% of families and 7.0% of the population were below the poverty line, including 10.0% of those under age 18 and 3.7% of those age 65 or over.

History

Sewickley Heights was established as a borough in 1935, but the area's character was largely established with the move of the Allegheny Country Club from Pittsburgh to its Sewickley Heights location in 1902. The establishment of the country club accelerated the settlement of the area as a haven for wealthy Pittsburgh residents. Many estates established in Sewickley Heights up through the 1930s occupied hundreds of acres with houses of immense proportions. Among the grandest estates was Thawmont, the estate of banker, shipper and investor William Thaw. Thawmont was featured in a 1903 print advertisement of the United States Battery Company that promoted electric lighting for country homes. Other notable estates included the Henry Robinson Rea mansion, Farmhill (which hosted Madame Curie in May 1921),[1] and B.F. Jones' 100-room mansion, Fairacres.

Many of the grand estates in Sewickley Heights began to fall into disrepair in the 1950s and 1960s. Many of the original massive houses were demolished and the lots subdivided. Sewickley Heights preserved the country character of the borough by requiring minimum lot sizes of 5 acres (m^2), though many homes are on substantially larger parcels. Many "neighborhoods" of Sewickley Heights are named after the original estate and the clusters of homes on the estate parcel are marked by unique stone fences original to the old estate.

In the 1960s and 1970s, several parcels of land were donated or purchased to form the Sewickley Heights Borough Park. The park now occupies approximately 600 acres (km^2) and is renowned regionally for its hiking and horse-riding trails and other recreation areas. Sewickley Heights is also home to the Fern Hollow Nature Center and the Sewickley Heights History Center, which are co-located on a 33-acre (m^2) site.

References

[1] http://news.google.com/newspapers?id=hEwbAAAAIBAJ&sjid=nEsEAAAAIBAJ&pg=2528%2C3948452

Allegheny_County,_Pennsylvania

Allegheny County is a county in the southwestern part of the U.S. state of Pennsylvania. As of the 2010 census, the population was 1,223,348; making it the second most populous county in Pennsylvania, following Philadelphia County. The county seat is Pittsburgh. The county forms the nucleus of the tri-state Pittsburgh metropolitan area and Pittsburgh DMA.

History

Allegheny County was the first in Pennsylvania to be given a Native American name, being named after the Allegheny River. The word "Allegheny" is of Lenape origin, with uncertain meaning. It is usually said to mean "fine river", but sometimes said to refer to an ancient mythical tribe called "Allegewi" who lived along the river long ago before being destroyed by the Lenape.[1]

Plaque on the statue next to the Frick Fine Arts Building

Not a great deal is known about the native inhabitants of the region prior to European contact. During the colonial era various native groups claimed or settled in the area, resulting in a multi-ethnic mix that included Iroquois, Lenape, Shawnee, and Mingo.

The first Europeans to enter the area were the French in 1749. Captain Pierre Joseph de Celeron, sieur de Blainville claimed the Ohio Valley and all of western Pennsylvania for Louis XV of France. The captain traveled along the Ohio and Allegheny rivers inserting lead plates in the ground to mark the land for France.

Since most of the towns during that era were developed along waterways, both the French and the British desired control over the local rivers. Therefore, the British sent Major George Washington to try to compel the French to leave their posts, with no success. Having failed in his mission, he returned and nearly drowned crossing the ice-filled Allegheny River. In 1754, the English tried again to enter the area. This time, they sent 41 Virginians to build Fort Prince George. The French got news of the plan and sent an army to take over the fort, which they then resumed building with increased fortification, renaming it Fort Duquesne.

The loss of the fort cost the English dearly because Fort Duquesne became one of the focal points of the French and Indian War. The first attempt to retake the fort, the Braddock Expedition, failed miserably.[2] It was not until General John Forbes attacked in 1758, four years after they had lost the original fort, that they recaptured and destroyed it. They subsequently built a new fort on the site, including a moat, and named it Fort Pitt. The site is now Pittsburgh's Point State Park.

Both Pennsylvania and Virginia claimed the region that is now Allegheny County. Pennsylvania administered most of the region as part of its Westmoreland County. Virginia considered everything south of the Ohio River and east of the Allegheny River to be part of its Yohogania County and governed it from Fort Dunmore. In addition, parts of the county were located in the proposed British colony of Vandalia and the proposed U.S. state of Westsylvania. The overlapping boundaries, multiple governments, and confused deed claims soon proved unworkable. In 1780

Pennsylvania and Virginia agreed to extend the Mason–Dixon line westward, and the region became part of Pennsylvania. From 1781 until 1788, much of what had been claimed as part of Yohogania County, Virginia, was administered as a part of the newly created Washington County, Pennsylvania.

Allegheny County was officially created on September 24, 1788, from parts of Washington and Westmoreland counties. It was formed due to pressure from settlers living in the area around Pittsburgh, which became the county seat in 1791. The county originally extended all the way north to the shores of Lake Erie and became the "mother county" for most of what is now northwestern Pennsylvania. By 1800, the county's current borders were set.

In the 1790s, a whiskey excise tax was imposed by the United States federal government. This started the so-called Whiskey Rebellion when the farmers who depended on whiskey income refused to pay and drove off tax collector John Neville. After a series of demonstrations by farmers, President George Washington sent troops to stop the rebellion.

The Allegheny County Courthouse

The area developed rapidly throughout the 19th century to become the center of steel production in the nation. Pittsburgh would later be labeled the "Steel Capital of the World".

Law and government

For most of the 20th century, until 1999, Allegheny County was governed exclusively under the state's Second Class County Code. Under this code, the county handled everything: elections, prisons, airports, public health and city planning. Unlike the rest of the state, where certain public offices are combined and held by one person, in Allegheny County all public offices are held by elected individuals.

Before the implementation of the home-rule charter on January 1, 2000, there were three county commissioners. These were replaced with an elected chief officer (the county executive), a county council with 15 members (13 elected by district, two elected county-wide), and an appointed county manager. The changes were intended to maintain a separation of powers between the executive and legislative branches while providing the citizens with greater control over the government.

County Medical Examiner office

The county has 130 municipalities, each governing itself; no other county in Pennsylvania has nearly as many, with Luzerne County's 76 being second.[3] The county has one Second Class City (Pittsburgh) and three Third Class Cities (Clairton, Duquesne, and McKeesport).

A 2004 study by the University of Pittsburgh stated that Allegheny County would be better served by consolidating the southeastern portion of the county (which includes many small, poor communities) into one large municipality, called "Rivers City," which would have a combined population of approximately 250,000.[4]

State relations

Under the Onorato administration, Allegheny County paid $10,000 per month to Robert Ewanco, of Greenlee Partners, to lobby the Pennsylvania General Assembly.[5] [] County officials credit him with a "20-fold" return in the form of appropriations for a widening project on Pennsylvania Route 28, as well as a footbridge and security cameras at Duquesne University.[]

County Executive

- Rich Fitzgerald, Democrat

County Council

- Matt Drozd, District 1, Republican
- Jan Rea, District 2, Republican
- Jim Burn, District 3, Democrat
- Michael J. Finnerty, District 4, Democrat
- Vince Gastgeb, District 5, Republican
- John F. Palmiere, District 6, Democrat
- Nicholas Futules, District 7, Democrat
- Charles Martoni, District 8, President, Democrat
- Robert J. Macey, District 9, Democrat
- William Russell Roberston, District 10, Democrat
- Barbara Daley-Danko, District 11, Democrat
- James Ellenbogen, District 12, Democrat
- Amanda Green Hawkins, District 13, Democrat
- John DeFazio, At-large, Democrat
- Edward J. Kress, At-large, Republican

Other elected county offices

- Controller, Chelsa Wagner, Democrat
- District Attorney, Stephen A. Zappala Jr., Democrat
- Sheriff, William P. Mullen, Democrat
- Treasurer, John K. Weinstein, Democrat

Politics

Presidential Election Results 1960–2012

Year	Democrat	Republican
2012	**56.54%** *352,687*	42.01% *269,039*
2008	**57.20%** *368,453*	41.89% *269,819*
2004	**57.15%** *368,912*	42.13% *271,925*
2000	**56.65%** *329,963*	40.41% *235,361*
1996	**52.82%** *284,480*	37.89% *204,067*
1992	**52.75%** *324,004*	29.80% *183,035*
1988	**59.51%** *348,814*	39.43% *231,137*

1984	**55.96%** *372,576*	42.76% *284,692*
1980	**47.87%** *297,464*	43.75% *271,850*
1976	**50.68%** *328,343*	46.79% *303,127*
1972	42.26% *282,496*	**55.60%** *371,737*
1968	**51.12%** *364,906*	37.09% *264,790*
1964	**66.03%** *475,207*	33.58% *241,707*
1960	**57.07%** *428,455*	42.76% *320,970*

As of November 2008, there are 955,982 registered voters in Allegheny County.[6]

- Democratic: 595,149 (62.26%)
- Republican: 256,291 (26.81%)
- Other Parties: 104,542 (10.94%)

The Republican Party had been historically dominant in county-level politics; prior to the Great Depression Pittsburgh and Allegheny County had been Republican. Since the Great Depression on the state and national levels, the Democratic Party has been dominant in county-level politics and is the Democratic stronghold of western Pennsylvania. In 2000, Democrat Al Gore won 56% of the vote and Republican George W. Bush won 41%. In 2004, Democrat John Kerry received 57% of the vote and Bush received 42%. In 2006, Democrats Governor Ed Rendell and Senator Bob Casey, Jr. won 59% and 65% of the vote in Allegheny County, respectively. In 2008, Democrat Barack Obama received 57% of the vote, John McCain received 41%, and each of the three state row office winners (Rob McCord for Treasurer, Jack Wagner for Auditor General, and Tom Corbett for Attorney General) also carried Allegheny.

State senators

- Matt Smith, Democrat, 37th district
- Jim Ferlo, Democrat, 38th district
- Randy Vulakovich, Republican, 40th district
- Wayne D. Fontana, Democrat, 42nd district
- Jay Costa, Democrat, 43rd district
- James Brewster, Democrat, 45th district
- Barry Stout, Democrat, 46th district
- Elder Vogel, Republican, 47th district

US representatives

- Keith Rothfus, Republican, 12th district
- Michael F. Doyle, Democrat, 14th district
- Tim Murphy, Republican, 18th district

Geography

Allegheny County is known for the three major rivers that flow through it: the Allegheny River and the Monongahela River converge at Downtown Pittsburgh to form the Ohio River. The Youghiogheny River flows into the Monongahela River at McKeesport, 10 miles (km) southeast. Several islands are located within the riverine systems. Water from these rivers eventually flows into the Gulf of Mexico via the Mississippi River. Although the county's industrial growth caused the clearcutting of forests, a significant woodland remains.

According to the U.S. Census Bureau, the county has a total area of 745 square miles (km^2); 730 square miles (km^2) is land and 15 square miles (km^2) (1.95%) is water.

Adjacent counties

- Butler County (north)
- Armstrong County (tangent to the northeast)
- Westmoreland County (east)
- Washington County (southwest)
- Beaver County (northwest)

Demographics

Historical populations		
Census	**Pop.**	**%±**
1790	10203	—
1800	15087	47.9%
1810	25317	67.8%
1820	34921	37.9%
1830	50552	44.8%
1840	81235	60.7%
1850	138290	70.2%
1860	178831	29.3%
1870	262204	46.6%
1880	355869	35.7%
1890	551959	55.1%
1900	775058	40.4%
1910	1018463	31.4%
1920	1185808	16.4%
1930	1374410	15.9%
1940	1411539	2.7%
1950	1515237	7.3%
1960	1628587	7.5%
1970	1605016	−1.4%
1980	1450085	−9.7%
1990	1336449	−7.8%
2000	1281666	−4.1%
2010	1223348	−4.6%
Est. 2012	1229866	0.5%
U.S. Decennial Census[7] 2011 estimate[8] through 1960[9]		

As of the census of 2010, there were 1,223,348 people residing in the county. The population density was 1676 people per square mile (647/km²). The racial makeup of the county was 82.87% White, 14.39% Black or African

American, 2.94% Asian, 0.03% Pacific Islander, 0.37% from other races, and 1.40% from two or more races. About 1.31% of the population were Hispanic or Latino of any race.

At the census[] of 2000, there were 1,281,666 people, 537,150 households, and 332,495 families residing in the county. The population density was 1,755 people per square mile (678/km²). There were 583,646 housing units at an average density of 799 per square mile (309/km²). The racial makeup of the county was 84.33% White, 12.41% Black or African American, 0.12% Native American, 1.69% Asian, 0.03% Pacific Islander, 0.34% from other races, and 1.07% from two or more races. About 0.87% of the population were Hispanic or Latino of any race. 20.0% were of German, 15.0% Italian, 12.7% Irish, 7.5% Polish and 5.1% English ancestry according to Census 2000. 93.5% spoke English and 1.3% Spanish as their first language.

There were 537,150 households out of which 26.40% had children under the age of 18 living with them, 46.10% were married couples living together, 12.40% had a female householder with no husband present, and 38.10% were non-families. Some 32.70% of all households were made up of individuals and 13.20% had someone living alone who was 65 years of age or older. The average household size was 2.31 and the average family size was 2.96.

The population was spread out with 21.90% under the age of 18, 8.50% from 18 to 24, 28.30% from 25 to 44, 23.40% from 45 to 64, and 17.80% who were 65 years of age or older. The median age was 40. For every 100 females, there were 90.00 males; for every 100 females age 18 and over, there were 86.20 males.

Economy

In the late 18th century farming played a critical role in the growth of the area. There was a surplus of grain due to transportation difficulties in linking with the eastern portion of the state. As a result, the farmers distilled the grain into whiskey, which significantly helped the farmers financially.

The area quickly became one of the key manufacturing areas in the young nation. Coupled with deposits of iron and coal, and the easy access to waterways for barge traffic, the city quickly became one of the most important steel producing areas in the world. Based on 2007 data from the US Army Corps of Engineers, Pittsburgh is the second busiest inland port in the nation.

With the decline of the steel industry in the US, the area shifted to other industries. Today, it is known for its hospitals, universities, and industrial centers. Despite the decline of heavy industry, Pittsburgh is home to a number of major companies and is ranked in the top ten [10] among US cities hosting headquarters of Fortune 500 corporations. These include U.S. Steel Corporation, PNC Financial Services Group, PPG Industries, and H. J. Heinz Company.

The county leads the commonwealth in number of defense contractors supplying the U.S. military.[11]

Regions

- North Hills
- South Hills
- West Hills
- City of Pittsburgh

Municipalities

Under Pennsylvania law, there are four types of incorporated municipalities: cities, boroughs, townships, and, in at most two cases, towns. Allegheny County also labels four of its municipalities as being simply "Municipalities" without using any of the specific qualifying labels.[] The following cities, municipalities, boroughs and townships are in Allegheny County:

Map of Allegheny County, Pennsylvania, with municipal labels showing cities and boroughs (red), townships (white), and census-designated places (blue)

Cities

- Clairton
- Duquesne
- McKeesport
- Pittsburgh

Municipalities

- Bethel Park
- Monroeville
- Mt. Lebanon
- Penn Hills

Boroughs

- Aspinwall
- Avalon
- Baldwin
- Bell Acres
- Bellevue
- Ben Avon
- Ben Avon Heights
- Blawnox
- Brackenridge
- Braddock
- Braddock Hills
- Bradford Woods
- Brentwood
- Bridgeville
- Carnegie
- Dormont
- Dravosburg
- East McKeesport
- East Pittsburgh
- Edgewood
- Edgeworth
- Elizabeth
- Emsworth
- Etna
- Forest Hills
- Fox Chapel
- Franklin Park
- Glassport
- Glen Osborne
- Glenfield
- Leetsdale
- Liberty
- Lincoln
- McDonald (mostly in Washington County)
- McKees Rocks
- Millvale
- Mount Oliver
- Munhall
- North Braddock
- Oakdale
- Oakmont
- Osborne (now Glen Osborne)
- Pennsbury Village
- Pitcairn
- Pleasant Hills
- Sewickley Heights
- Sewickley Hills
- Sharpsburg
- Springdale
- Swissvale
- Tarentum
- Thornburg
- Trafford (mostly in Westmoreland County)
- Turtle Creek
- Verona
- Versailles
- Wall
- West Elizabeth
- West Homestead
- West Mifflin

- Castle Shannon
- Chalfant
- Cheswick
- Churchill
- Coraopolis
- Crafton
- Green Tree
- Haysville
- Heidelberg
- Homestead
- Ingram
- Jefferson Hills
- Plum
- Port Vue
- Rankin
- Rosslyn Farms
- Sewickley
- West View
- Whitaker
- White Oak
- Whitehall
- Wilkinsburg
- Wilmerding

Townships

- Aleppo Township
- Baldwin Township
- Collier Township
- Crescent Township
- East Deer Township
- Elizabeth Township
- Fawn Township
- Findlay Township
- Forward Township
- Frazer Township
- Hampton Township
- Harmar Township
- Harrison Township
- Indiana Township
- Kennedy Township
- Kilbuck Township
- Leet Township
- Marshall Township
- McCandless
- Moon Township
- Neville Township
- North Fayette Township
- North Versailles Township
- O'Hara Township
- Ohio Township
- Pine Township
- Reserve Township
- Richland Township
- Robinson Township
- Ross Township
- Scott Township
- Shaler Township
- South Fayette Township
- South Park Township
- South Versailles Township
- Springdale Township
- Stowe Township
- Upper St. Clair
- West Deer Township
- Wilkins Township

Census-designated places

Census-designated places are geographical areas designated by the US Census Bureau for the purposes of compiling demographic data. They are not actual jurisdictions under Pennsylvania law. Other unincorporated communities, such as villages, may be listed here as well.

- Allison Park
- Bairdford
- Bakerstown
- Boston
- Carnot-Moon
- Clinton
- Curtisville
- Enlow
- Gibsonia
- Glenshaw
- Greenock
- Harwick
- Imperial
- Noblestown
- Rennerdale []
- Russellton
- Sturgeon

Unincorporated communities

- Acmetonia
- Blackridge
- Blanchard
- Broughton
- Bruceton
- Creighton
- Cubbage Hill []
- Dorseyville
- Ewingsville []
- Harmarville
- Hickory Heights
- Ingomar
- Karns
- Keown Station
- Kirwan Heights []
- Library
- Mount Vernon
- Natrona
- Natrona Heights
- Nevillewood []
- Presto []
- Regent Square
- Wexford
- Wildwood

Former municipalities and political subdivisions

Many political subdivisions of Allegheny County have come and gone through subdivision or annexation through the years. These include:

- Allegheny City – the area that is now the North Shore (or North Side) of the City of Pittsburgh, north of the Allegheny River.
- Allentown Borough – now the neighborhood of Allentown in Pittsburgh.
- Birmingham Borough – what is now Pittsburgh's South Side.
- Carrick Borough – now the neighborhood of Carrick. Formed out of Baldwin Township in 1904, this borough existed until it was annexed by Pittsburgh in 1927. It was named for Carrick-on-Suir, Ireland. To this day, some of the manhole covers bear the Carrick Borough name.
- Collins Township – in what is now the Strip District and Lawrenceville of the City of Pittsburgh.
- McClure Township – McClure was formed in 1858 from the section of Ross Township adjacent to Allegheny City. In 1867 McClure, along with sections of Reserve Township, was incorporated into Allegheny City. The McClure section of this annexation became Wards 9 (Woods Run Area) and 11 (present day Brighton Heights) in the City of Pittsburgh.
- Patton Township – was in east central part of the county, north of North Versailles Township, east of Wilkins and Penn Townships, and south of Plum Township. In U.S. census for 1860–1880.
- Peebles Township – included most of what is now the eastern part of the city of Pittsburgh from the Monongahela River in the south (today's Hazelwood) to the Allegheny River in the north. It was subdivided into Collins and Liberty townships, all of which were incorporated into Pittsburgh in 1868.
- St. Clair Township – stretched from the Monongahela River south to the Washington County line. It divided into Lower St. Clair, which eventually became part of the City of Pittsburgh, and Upper St. Clair.
- Snowden – now known as South Park Township.
- Temperanceville – what is now Pittsburgh's West End.
- Union Borough – the area surrounding Temperanceville.

Education

Colleges and universities

Full list of colleges and universities in Pittsburgh

- The Art Institute of Pittsburgh
- Byzantine Catholic Seminary of SS. Cyril and Methodius
- Carlow University
- Carnegie Mellon University
- Chatham University
- DeVry University
- Duquesne University

- La Roche College
- Penn State Greater Allegheny
- Pittsburgh Theological Seminary
- Point Park University
- Reformed Presbyterian Theological Seminary
- Robert Morris University
- University of Pittsburgh

Community, junior and technical colleges

- Bidwell Training Center
- Bradford School
- Career Training Academy
- Community College of Allegheny County
- Dean Institute of Technology
- Duff's Business Institute
- ICM School of Business and Medical Careers
- International Academy of Design and Technology
- ITT Technical Institute
- Le Cordon Bleu Institute of Culinary Arts in Pittsburgh
- Median School of Allied Health Careers
- Pittsburgh Beauty Academy
- Pittsburgh Institute of Aeronautics
- Pittsburgh Institute of Mortuary Science
- Pittsburgh Technical Institute
- Rosedale Technical Institute
- Triangle Tech
- Western School of Health and Business Careers

Public school districts

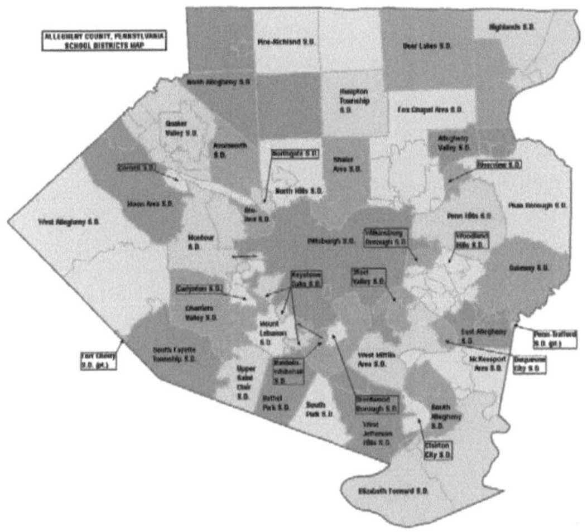

Map of Allegheny County, Pennsylvania Public School Districts

- Allegheny Valley School District
- Avonworth School District
- Baldwin-Whitehall School District
- Bethel Park School District
- Brentwood Borough School District
- Carlynton School District
- Chartiers Valley School District
- Clairton City School District
- Cornell School District
- Deer Lakes School District
- Duquesne City School District
- East Allegheny School District
- Elizabeth Forward School District
- Fort Cherry School District (part)
- Fox Chapel Area School District
- Gateway School District
- Hampton Township School District
- Highlands School District
- Keystone Oaks School District
- McKeesport Area School District
- Montour School District
- Moon Area School District
- Mt. Lebanon School District
- North Allegheny School District
- North Hills School District
- Northgate School District
- Penn Hills School District
- Penn-Trafford School District (part)
- Pine-Richland School District
- Pittsburgh School District
- Plum Borough School District
- Quaker Valley School District
- Riverview School District
- Shaler Area School District
- South Allegheny School District
- South Fayette Township School District
- South Park School District
- Steel Valley School District
- Sto-Rox School District
- Upper St. Clair School District
- West Allegheny School District
- West Jefferson Hills School District
- West Mifflin Area School District
- Wilkinsburg School District
- Woodland Hills School District

Public charter schools

Pennsylvania charter schools participate in PSSA testing just like all public schools.

- Academy Charter School - Pittsburgh
- Career Connections Charter HIgh School - Pittsburgh
- City Charter School - Pittsburgh
- Environmental Charter School at Frick Park – 3rd grade
- Manchester Academic Charter School – Liverpool Street, Pittsburgh
- MT Lebanon Montessori School and Academy PreK age 3 through 6th grade
- Northside Urban Pathways Charter School 6th–12th grades - Pittsburgh
- PA Learners Online Regional Charter School (public cyber school) - Homestead
- Penn Hills Charter School for Entrepreneurship - Verona
- Pennsylvania Distance Learning Charter School - Wexford
- Propel Braddock Hills CS - Pittsburgh
- Propel East – Turtle Creek
- Propel Homestead – Homestead
- Propel McKeesport - McKeesport
- Propel Northside – Pittsburgh
- Propel Montour – Kennedy Township
- Propel Andrew Street High School – Munhall
- Spectrum Charter School – 7th–12th grades Monroeville
- Urban League of Greater Pittsburgh Charter School – Kindergarten-5th grades Wood Street, Pittsburgh
- Urban Pathways K-5 College Charter School
- Young Scholars of Western Pennsylvania CS

per 2012 - Pennsylvania Department of Education EdNA Allegheny County, Pennsylvania

Approved private schools and charter schools for the blind and deaf

The Commonwealth of Pennsylvania has 36 Approved Private Schools including the Charter Schools for the Blind and Deaf. Students attending these schools come from across the commonwealth. The private schools are licensed by the State Board of Private Academic Schools. They provide a free appropriate special education for students with severe disabilities. The cost of tuition for these schools is paid 60% by the state and 40% by the local school district where the student is a resident. Pennsylvania currently has four PA chartered and 30 non-charter APSs for which the Department approves funding. These schools provide a program of special education for over 4,000 day and residential students. Parents are not charged for the services at the school.[12] In 2009, the Pennsylvania Department of Education budgeted $98 million for tuition of children in approved private schools and $36.8 million for students attending the charter schools for the deaf and blind.[13] The majority of these schools are located in the southeastern region and southwestern region of Pennsylvania.

- ACLD Tillotson School, Pittsburgh – Tuition rate $38,804
- The Day School at The Children's Institute, Pittsburgh – Tuition rate $55,217
- DePaul School for Hearing and Speech, Pittsburgh – Tuition rate $36,892
- Easter Seal Society of Western Pennsylvania – Tuition rate $60,891.97
- The Education Center at the Watson Institute, Sewickley – Tuition rate $42,242
- Pace School, Pittsburgh – Tuition rate – $37,635
- Pressley Ridge Day School, Pittsburgh – Tuition rate – $51,177
- Pressley Ridge School for the Deaf, Pittsburgh – Tuition rate – $66,022, residential $128,376
- The Watson Institute Friendship Academy, Pittsburgh – Tuition rate – $38,211
- Wesley Spectrum Highland Services, Pittsburgh – Tuition rate – $39,031

- Western Pennsylvania School for Blind Children, Pittsburgh – Tuition rate – $82,500, residential $120,100
- Western Pennsylvania School for the Deaf, Pittsburgh – Tuition rate – $61,051, residential – $99,919

Private high schools

- Bishop Canevin High School
- Central Catholic High School
- Mount Alvernia High School
- North Catholic High School
- Oakland Catholic High School
- Our Lady of the Sacred Heart High School
- Serra Catholic High School
- Seton-La Salle Catholic High School
- Sewickley Academy
- Shady Side Academy
- St. Joseph High School
- Vincentian Academy

21st Century Community Learning Centers

These are state designated before and after school program providers. They receive state funding through grants. CCLCs provide academic, artistic and cultural enhancement activities to students and their families when school is not in session.[14]

- Boys & Girls Clubs of Western PA – 2010 Grant – $261,748
- Cornell School District – 2010 Grant – $526,800
- Human Services Center Corporation – 2010 Grant- $550,000
- McKeesport Area School District – 2010 Grant – $468,000
- Penn Hills School District – 2010 Grant – $360,000
- The Hill House/One Small Step –2010 Grant – $675,000
- Wireless Neighborhoods – 2010 Grant – $612,000

Transportation

Allegheny County's public transportation provider is the Port Authority of Allegheny County. The Allegheny County Department of Public Works oversees infrastructure, maintenance and engineering services in the county.

Major roadways

- Interstate 79 runs north to south from Wexford to Bridgeville
- Interstate 279 runs north to south from the North Hills to Downtown
- Interstate 579 (Crosstown Boulevard)
- Interstate 76 / PA Turnpike runs east to west from I-376 in Monroeville to the Warrendale interchange (near I-79)
- Interstate 376 runs east to west from the eastern suburbs (Monroeville, Plum, Penn Hills) across the county to Pittsburgh International Airport and beyond
- Pennsylvania Turnpike 576 (future I-576) (Findlay Connector)
- US Route 19 runs north to south from Wexford to Mt. Lebanon
- US Route 22 runs west to east and makes up US Route 30 and I-376
- US Route 30 runs west to east and joins US 22 and I-376 near Pittsburgh International Airport

For information about major state roads, see list of State Routes in Allegheny County, Pennsylvania and Allegheny County Belt System.

Parks and recreation

There are two Pennsylvania state parks in Allegheny County. Point State Park is at the confluence of the Allegheny and Monongahela rivers in Downtown Pittsburgh, and Allegheny Islands State Park is in the Allegheny River in Harmar Township and is undeveloped as of August 2010.

Notable people

Further information: List of people from the Pittsburgh metropolitan area

Major League Sports teams

- Pittsburgh Steelers, football team
- Pittsburgh Penguins, ice hockey team
- Pittsburgh Pirates, baseball team
- Pittsburgh Power, arena football team

See also

- List of Pennsylvania state historical markers in Allegheny County
- National Register of Historic Places listings in Allegheny County, Pennsylvania
- Spanish–American War
- The ScareHouse

References

[2] Fiske, John (1902). *New France and New England* (http://books.google.com/books?id=DiATAAAAYAAJ&pg=PA290), pp. 290–92. Houghton Mifflin Company.

[10] http://www.downtownpittsburgh.com/doingBusinessDowntown.aspx

[12] Approved Private Schools and Chartered Schools for the Deaf and the Blind, Pennsylvania Department of Education website, accessed April 2010.

[13] Tommasini, John, Assistant Secretary of Education, Testimony before the Pennsylvania Senate Education Committee Hearing on SB982 of 2010. given April 14, 2010.

[14] Pennsylvania Awards $29.9 Million to Support 21st Century Community Learning Centers, Pennsylvania Department of Education Press Release, April 7, 2010

External links

- Allegheny County official website (http://www.alleghenycounty.us)
- Allegheny County Quest (http://alleghenycounty.wikidot.com)
- County Map by Municipality (http://www.alleghenycounty.us/munimap/index.asp)
- Historic Pittsburgh Map Collection (http://digital.library.pitt.edu/maps/)
- Southwest Pennsylvania Commission (http://www.spcregion.org/reg_all.shtml)

Pennsylvania

Pennsylvania /ˌpɛnsɪlˈveɪnjə/, officially the **Commonwealth of Pennsylvania**, is a US state that is located in the Northeastern and Mid-Atlantic regions of the United States, and the Great Lakes region. The state borders Delaware to the southeast, Maryland to the south, West Virginia to the southwest, Ohio to the west, Lake Erie and Ontario, Canada to the northwest, New York to the north and New Jersey to the east. The Appalachian Mountains run through the middle of the state.

Pennsylvania is the 33rd most extensive, the 6th most populous, and the 9th most densely populated of the 50 United States. The state's four most populous cities are Philadelphia, Pittsburgh, Allentown, and Erie. The state capital is Harrisburg. Pennsylvania has 51 miles (km) of coastline along Lake Erie and 57 miles (km)[] of shoreline along the Delaware Estuary. The state is one of the 13 original founding states of the U.S.

Geography

Further information: Geography of Pennsylvania and List of counties in Pennsylvania

Pennsylvania is 170 miles (km) north to south and 283 miles (km) east to west.[] Of a total 46055 square miles (km^2), 44817 square miles (km^2) are land, 490 square miles (km^2) are inland waters, and 749 square miles (km^2) are waters in Lake Erie.[1] It is the 33rd largest state in the United States.[2] Pennsylvania has 51 miles (km)[] of coastline along Lake Erie and 57 miles (km)[] of shoreline along the Delaware Estuary.

Counties of Pennsylvania.

The bounds of the state are the Mason–Dixon line (39° 43' N) to the south, the Twelve-Mile Circle on the Pennsylvania-Delaware border, the Delaware River to the east, 80° 31' W to the west and the 42° N to the north, with the exception of a short segment on the western end, where a triangle extends north to Lake Erie.

Pennsylvania borders six other states: New York to the north; New Jersey to the east; Delaware and Maryland to the southeast; West Virginia to the southwest, and Ohio to the west. Pennsylvania also shares a water border with the Canadian province of Ontario. Of the original Thirteen

Colonies, Pennsylvania is the only state that does not border the Atlantic Ocean.

It has the cities of Philadelphia, Reading, Lebanon and Lancaster in the southeast, Pittsburgh in the southwest, the tri-cities of Allentown, Bethlehem, and Easton in the central east (known as the Lehigh Valley), the tri-cities of Scranton, Wilkes-Barre, and Hazleton in the northeast, and Erie in the northwest. Williamsport serves as the "hub" of the commonwealth's north-central region, with York and the state capital Harrisburg on the Susquehanna River in the central region of the commonwealth.

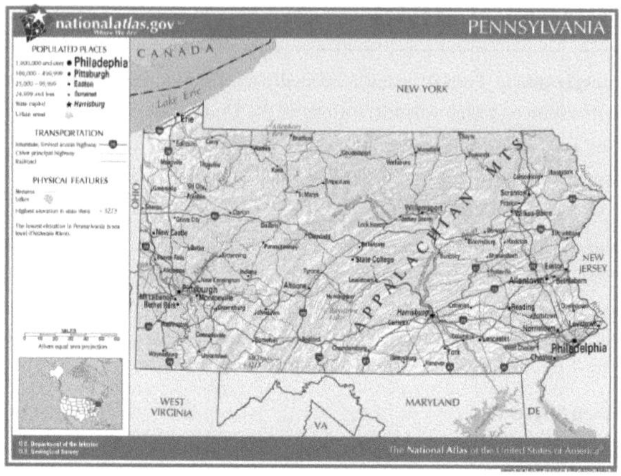

Map of Pennsylvania, showing major cities and roads.

There are 5 regions. They are the Allegheny Plateau, Ridge and Valley, Atlantic Coastal Plain, Piedmont, and the Erie Plain.

Climate

Pennsylvania's diverse topography also produces a variety of climates, though the entire state experiences cold winters and humid summers. Straddling two major zones, the majority of the state, with the exception of the southeastern corner, has a humid continental climate (Köppen climate classification *Dfa*). Greater Philadelphia has some characteristics of the humid subtropical climate (Köppen *Cfa*) that covers much of Delaware and Maryland to the south.

Moving toward the mountainous interior of the state, the winter climate becomes markedly colder, the number of cloudy days increases, and snowfall amounts are greater. Western areas of the state, particularly locations near Lake Erie, can receive over 100 inches (cm) of snowfall annually, and the entire state receives plentiful precipitation throughout the year. The state may be subject to severe weather from spring through summer into fall; an average of 10 tornadoes touch down each year in the state.

Monthly Normal High and Low Temperatures For Various Pennsylvania Cities in °F												
City	**Jan.**	**Feb.**	**Mar.**	**Apr.**	**May.**	**Jun.**	**Jul.**	**Aug.**	**Sep.**	**Oct.**	**Nov.**	**Dec.**
Scranton	33/19	37/21	46/28	59/38	70/48	78/56	82/61	80/60	72/52	61/41	49/33	38/24
Erie	34/21	36/21	44/27	56/38	67/48	76/58	80/63	79/62	72/56	61/45	50/37	38/27
Pittsburgh	36/21	39/23	49/30	62/40	71/49	79/58	83/63	81/62	74/54	63/43	51/35	39/25

Harrisburg	37/23	41/25	50/33	62/42	72/52	81/62	85/66	83/64	76/56	64/45	53/35	41/27
Philadelphia	40/26	44/28	53/34	64/44	74/54	83/64	87/69	85/68	78/60	67/48	56/39	45/30
Allentown	36/20	40/22	49/29	61/39	72/48	80/58	84/63	82/61	75/53	64/41	52/33	40/24
[3], [4], [5], [6], [7],												

History

Before the Commonwealth was settled by Europeans, the area was home to the Delaware (also known as Lenni Lenape), Susquehannock, Iroquois, Eriez, Shawnee, and other American Indian Nations.[8] Both the Dutch and the English claimed both sides of the Delaware River as part of their colonial lands in America.[9] [10] [11] The Dutch were the first to take possession, which has impact on the history of Pennsylvania.[12]

By June 3, 1631, the Dutch had started the settlement of the Delmarva Peninsula by establishing the Zwaanendael Colony on the site of present day Lewes, Delaware.[13] In 1638, Sweden heated up the issue by establishing the New Sweden Colony, centered on Fort Christina, on the site of present day Wilmington, Delaware. New Sweden claimed and, for the most part, controlled the lower Delaware River region (Parts of present Delaware, New Jersey and Pennsylvania) but settled few colonists there.[14] [15]

On March 12, 1664, King Charles II of England gave James, Duke of York a grant that included all of the lands included in the original Virginia Company of Plymouth Grant as well as other lands. This grant was – again – in conflict with the Dutch claim for New Netherland, which included parts of today's Pennsylvania.[16]

Penn's Treaty with the Indians, by Edward Hicks.

On June 24, 1664, The Duke of York sold the portion of his large grant that included present day New Jersey to John Berkeley and George Carteret for a proprietary colony. The land was not yet in British possession, but the sale boxed in the portion of New Netherland on the West side of the Delaware River. The British conquest of New Netherland was commenced on August 29, 1664, when New Amsterdam was coerced to surrender, facing the cannons on British ships in New York Harbor.[17] [18] This conquest continued, and was completed in October 1664, when the British captured Fort Casimir in what today is New Castle, Delaware.

The Peace of Breda between England, France and the Netherlands confirmed the British conquest on July 21, 1667,[19] [20] although there were temporary reversions.

On September 12, 1672, as part of the Third Anglo—Dutch War, the Dutch re-conquered New York Colony/New Amsterdam, the Dutch established three County Courts which went on to become original Counties in present day Delaware and Pennsylvania. The one that later transferred to Pennsylvania was Upland.[21] This was partially reversed on February 9, 1674, when the Treaty of Westminster ended the Third Anglo-Dutch War, and reverted all political situations to the *status quo ante bellum*. The British retained the Dutch Counties with their Dutch names.[22] By June 11, 1674, New York reasserted control over the outlying colonies, including Upland, but the names started to be changed to British names by November 11, 1674.[23] Upland was partitioned on November 12, 1674, producing the general outline of the current border between Pennsylvania and Delaware.[24]

John Dickinson

On February 28, 1681, Charles II granted a land charter[25] to William Penn to repay a debt of £16,000[26] (around £2,100,000 in 2008, adjusting for retail inflation)[27] owed to William's father, Admiral William Penn. This was one of the largest land grants to an individual in history.[] It was called Pennsylvania. William Penn, who wanted it called New Wales or Sylvania, was embarrassed at the change, fearing that people would think he had named it after himself, but King Charles would not rename the grant.[28] Penn established a government with two innovations that were much copied in the New World: the county commission and freedom of religious conviction.[]

What had been Upland on what became the Pennsylvania side of the Pennsylvania-Delaware Border was renamed as Chester County when Pennsylvania instituted their colonial governments on March 4, 1681.[29] [30] The Quaker leader William Penn had signed a peace treaty with Tammany, leader of the Delaware tribe, beginning a long period of friendly relations between the Quakers and the Indians.[31] Additional treaties between Quakers and other tribes followed. The treaty of William Penn was never violated.[32]

18th century

Between 1730 and when it was shut down by Parliament with the Currency Act of 1764, the Pennsylvania Colony made its own paper money to account for the shortage of actual gold and silver. The paper money was called Colonial Scrip. The Colony issued "bills of credit", which were as good as gold or silver coins because of their legal tender status. Since they were issued by the government and not a banking institution, it was an interest-free proposition, largely defraying the expense of the government and therefore taxation of the people. It also promoted general employment and prosperity, since the Government used discretion and did not issue too much to inflate the currency. Benjamin Franklin had a hand in creating this currency, of which he said its utility was never to be disputed, and it also met with the "cautious approval" of Adam Smith.[33]

After the Stamp Act Congress of 1765, Delegate John Dickinson of Philadelphia, Pennsylvania, wrote the *Declaration of Rights and Grievances*. The Congress was the first meeting of the thirteen colonies, called at the request of the Massachusetts Assembly, but only nine colonies sent delegates.[34] Dickinson then wrote *Letters from a Farmer in Pennsylvania, To the Inhabitants of the British Colonies*, which were published in the Pennsylvania Chronicle between December 2, 1767, and February 15, 1768.[35]

President's House, Philadelphia. The Masters-Penn mansion housed Pennsylvania's governor in the early 1770s. It later served as the presidential mansion of George Washington and John Adams, 1790–1800, while Philadelphia was the temporary national capital.

When the Founding Fathers of the United States convened in Philadelphia in 1774, 12 colonies sent representatives to the First Continental Congress.[36] The Second Continental Congress, which also met in Philadelphia (in May, 1775), drew up and signed the Declaration of Independence in Philadelphia,[37] but when that city was captured by the British, the Continental Congress escaped westward, meeting at the Lancaster courthouse on Saturday, September 27, 1777, and then to York. There they drew up the Articles of Confederation that formed 13 independent colonies into a new nation. Later, the Constitution was written, and Philadelphia was once again chosen to be cradle to the new American Nation.[38]

Pennsylvania became the second state to ratify the U.S. Constitution on December 12, 1787,[39] five days after Delaware became the first.

Dickinson College of Carlisle was the first college founded in the United States. Established in 1773, the college was ratified five days after the Treaty of Paris on September 9, 1783. The school was founded by Benjamin Rush and named after John Dickinson.

For half a century, the Commonwealth's General Assembly (legislature) met at various places in the general Philadelphia area before starting to meet regularly in Independence Hall in Philadelphia for 63 years.[] But it needed a more central location, as for example the Paxton Boys massacres of 1763 had made the legislature aware. So, in 1799 the General Assembly moved to the Lancaster Courthouse,[] and finally in 1812 to Harrisburg.[]

The "Hills Capitol", used from 1821 until it burned down in 1897.

The General Assembly met in the old Dauphin County Court House until December 1821,[] when the Federal-style "Hills Capitol" (named for its builder, Stephen Hills, a Lancaster architect) was constructed on a hilltop land grant of four acres set aside for a seat of state government by the prescient, entrepreneurial son and namesake of John Harris, Sr., a Yorkshire native who had founded a trading post in 1705 and ferry (1733) on the east shore of the Susquehanna River.[40] The Hills Capitol burned down on February 2, 1897, during a heavy snowstorm, presumably because of a faulty flue.[] The General Assembly met at Grace Methodist Church on State Street (still standing) until the a new capitol could be built. Following an architectural selection contest that many alleged had been "rigged," Chicago architect Henry Ives Cobb was charged with designing and building a replacement building; however, the legislature had little money to allocate to the project, and a roughly finished, somewhat industrial building (the Cobb Capitol) was completed. The General Assembly refused to occupy the building. Political and popular indignation in 1901 prompted a second contest that was restricted to Pennsylvania architects, and Joseph Miller Huston of Philadelphia was chosen to design the present Pennsylvania State Capitol that incorporated Cobb's building into magnificent public work finished and dedicated in 1907.[]

The new state Capitol drew rave reviews.[] Its dome was inspired by the domes of St. Peter's Basilica in Rome and the United States Capitol.[] President Theodore Roosevelt called it "the most beautiful state Capital in the nation" and said, "It's the handsomest building I ever saw" at the dedication. In 1989, the *New York Times* praised it as "grand, even awesome at moments, but it is also a working building, accessible to citizens ... a building that connects with the reality of daily life".[]

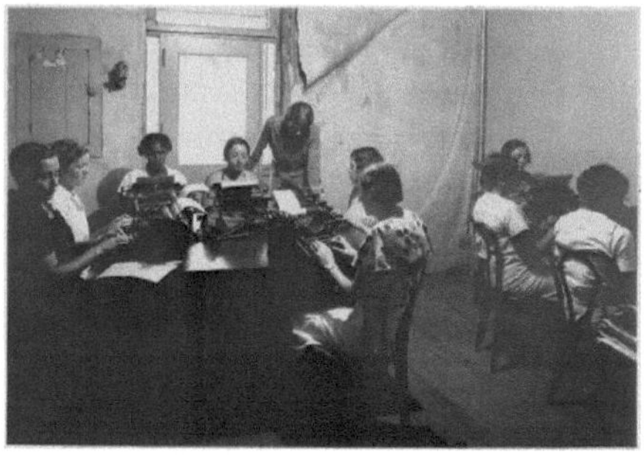

Roosevelt's FERA camp for unemployed women, 1934

Pennsylvania accounts for nine percent of wooded areas in the United States. In 1923 President Calvin Coolidge established the Allegheny National Forest under the authority of the Weeks Act of 1911 in the northwest part of the state in Elk, Forest, McKean, and Warren Counties for the purposes of timber production and watershed protection in the Allegheny River basin. The Allegheny is the state's only national forest.

James Buchanan, of Franklin County, the only bachelor President of the United States,[] was the only one to be born in Pennsylvania. The Battle of Gettysburg—the major turning point of the Civil War—took place near Gettysburg.[41] An estimated 350,000 Pennsylvanians served in the Union Army forces along with 8,600 African American military volunteers.

Pennsylvania was also the home of the first commercially drilled oil well. In 1859, near Titusville, Pennsylvania, Edwin L. Drake successfully drilled the well, which led to the first major oil boom in United States history.

Demographics

Further information: List of people from Pennsylvania

The United States Census Bureau estimates that the population of Pennsylvania was 12,763,536 on July 1, 2012, a 0.5% increase since the 2010 United States Census.[]

Of the people residing in Pennsylvania, 74.5% were born in Pennsylvania, 18.4% were born in a different US state, 1.5% were born Puerto Rico, U.S. Island areas, or born abroad to American parent(s), and 5.6% were foreign born.[42]

According to the 2010 Census, 81.9% of the population was White (79.2% non-Hispanic white), 11.3% was Black or African American, 0.3% American Indian and Alaska Native, 2.9% Asian, 1.9% from two or more races. 5.9% of the total population was of Hispanic or Latino origin (they may be of any race).[43]

As of 2011, 32.1% of Pennsylvania's population younger than age 1 were minorities.[44]

Pennsylvania's Hispanic population grew by 82.6% between 2000 and 2010, making it one of the largest increases in a state's Hispanic population. The significant growth of the Hispanic population is due to immigration to the state mainly from Puerto Rico, which is a US territory, but to a lesser extent from countries such as the Dominican Republic, Mexico, and various Central and South American nations, as well as from the wave of Hispanics leaving New York and New Jersey for safer and more affordable living. The Asian population swelled by almost 60%, which was fueled by Indian, Vietnamese, and Chinese immigration, as well the many Asian transplants moving to Philadelphia from New York. The rapid growth of this community has given Pennsylvania one of the largest Asian populations in the nation by numerical values. The Black and African American population grew by 13%, which was the largest increase in that population amongst the state's peers (New York, New Jersey, Ohio, Illinois, and Michigan).The White population declined by 0.7%, a trend that is beginning to reverse itself. Twelve other states saw decreases in their White populations.[45]

The center of population of Pennsylvania is located in Perry County, in the borough of Duncannon.[46]

As of 2006, Pennsylvania has an estimated population of 12,440,621, which is an increase of 35,273 from the previous year and an increase of 159,567 since the year 2000. Net migration to other states resulted in a decrease of 27,718, and immigration from other countries resulted in an increase of 127,007. Net migration to the Commonwealth was 98,289. Migration of native Pennsylvanians resulted in a decrease of 100,000 people. In 2006, 5.00% of Pennsylvanians were foreign born (621,480 people). The state has an estimated 2005 poverty rate of 11.9%.[] The state also has the 3rd highest proportion of elderly (65+) citizens in 2005.[]

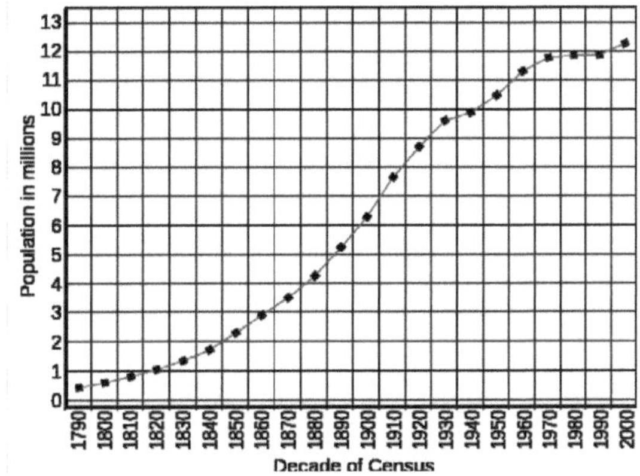

State population from 1790 to 2000

Foreign born Pennsylvanians are largely from Asia (36.0%), Europe (35.9%), Latin America (30.6%), Africa (5%), North America (3.1%), and Oceania (0.4%).

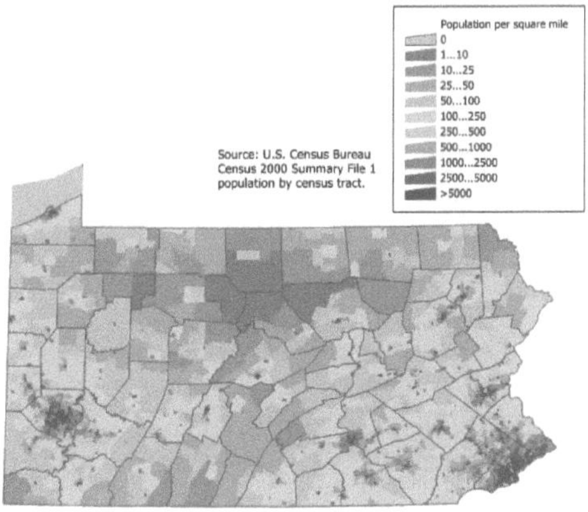

Pennsylvania's population distribution.

Pennsylvania's reported population of Hispanics, especially among the Asian, Hawaiian and White races, has markedly increased in recent years.[] The Hispanic population is greatest in Allentown, Lancaster, Reading, Hazleton, and around Philadelphia, with over 20% being Hispanic. It is not clear how much of this change reflects a changing population and how much reflects increased willingness to self-identify minority status. As of 2010, it is estimated that about 85% of all Hispanics in Pennsylvania live within a 150 miles (km) radius of Philadelphia, with about 20% living within the city itself.

Pennsylvania's population was reported as 5.9% under 5 and 23.8% under 18, with 15.6% aged 65 or older. Females made up 51.7% of the population.[] The largest ancestry groups are listed below, expressed as a percentage of total people who responded with a particular ancestry for the 2006–2008 census:[47]

- 28.5% German
- 18.2% Irish

- 12.8% Italian
- 9.6% African
- 8.5% English
- 7.2% Polish
- 4.2% French Canadian
- 2.9% Puerto Rican[]
- 2.2% Dutch
- 2.0% Slovak
- 2.0% Scotch Irish
- 1.7% Scottish
- 1.6% Russian
- 1.5% Welsh
- 1.2% Hungarian
- 1.0% West Indian
- 1.0% Ukrainian
- 1.0% Mexican

The five largest estimated ancestry groups in Pennsylvania are: German (28.5%), Irish (18.2%), Italian (12.8%), English (8.5%) and Polish (7.2%).

Demographics of Pennsylvania (csv) [48]

By race	White	Black	AIAN*	Asian	NHPI*
2000 (total population)	87.60%	10.71%	0.43%	2.04%	0.07%
2000 (Hispanic only)	2.74%	0.44%	0.06%	0.03%	0.02%
2005 (total population)	86.83%	11.20%	0.45%	2.46%	0.09%
2005 (Hispanic only)	3.52%	0.53%	0.07%	0.05%	0.02%
Growth 2000–05 (total population)	0.32%	5.83%	5.64%	22.23%	18.99%
Growth 2000–05 (non-Hispanic only)	-0.64%	5.21%	2.77%	21.86%	14.13%
Growth 2000–05 (Hispanic only)	29.86%	20.24%	23.61%	45.64%	35.44%
* AIAN is American Indian or Alaskan Native; NHPI is Native Hawaiian or Pacific Islander					

Historical populations

Census	Pop.	%±
1790	434373	—
1800	602365	38.7%
1810	810091	34.5%
1820	1049458	29.5%
1830	1348233	28.5%
1840	1724033	27.9%
1850	2311786	34.1%
1860	2906215	25.7%
1870	3521951	21.2%
1880	4282891	21.6%

1890	5258113	22.8%
1900	6302115	19.9%
1910	7665111	21.6%
1920	8720017	13.8%
1930	9631350	10.5%
1940	9900180	2.8%
1950	10498012	6.0%
1960	11319366	7.8%
1970	11793909	4.2%
1980	11863895	0.6%
1990	11881643	0.1%
2000	12281054	3.4%
2010	12702379	3.4%
Est. 2012	12763536	0.5%
Source: 1910–2010[49]		

Languages

Top 10 Non-English Languages Spoken in Pennsylvania

Language	Percentage of population (as of 2010)[]
Spanish	4.09%
Chinese (including Mandarin) and German (tied)	0.47%
Italian	0.43%
Pennsylvania Dutch	0.40%
French	0.34%
Russian and Vietnamese (tied)	0.29%
Korean	0.25%
Polish	0.21%
Arabic	0.20%
Hindi	0.17%

As of 2010, 90.15% (10,710,239) of Pennsylvania residents age 5 and older spoke English at home as a primary language, while 4.09% (486,058) spoke Spanish, 0.47% (56,052) Chinese (which includes Mandarin,) and German was spoken as a main language by 0.47% (55,891) of the population over the age of five. In total, 9.85% (1,170,628) of California's population age 5 and older spoke a mother language other than English.[]

Pennsylvania Dutch

The term "Dutch," when referring to the language spoken by the Pennsylvania Dutch people, (Pennsylvania Dutch), means "German" or "Teutonic" rather than "Netherlander." Germans, in their own language, call themselves "Deutsch," which in English became, misleadingly, "Dutch."[50] The Pennsylvania Dutch language is a descendant of German, in the West Central German dialect family. Although it is still spoken as a first language among some Old Order Amish and Mennonites (principally in the Lancaster County area), the language is almost extinct as an everyday language among the non-religious, though a few words have passed into English usage.

Religion

> “The new sovereign also enacted several wise and wholesome laws for his colony, which have remained invariably the same to this day. The chief is, to ill–treat no person on account of religion, and to consider as brethren all those who believe in one God.”
> – Voltaire, speaking of William Penn[51]

Of all the colonies, only Rhode Island had religious freedom as secure as in Pennsylvania, and one result was a wide religious diversity, which continues to this day.[52]

Pennsylvania's population in 2000 was 12,281,054. Of these, 8,448,193 were estimated to belong to some sort of organized religion. According to the Association of religion data archives at Pennsylvania State University, reliable data exists for 7,116,348 religious adherents in Pennsylvania in 2000 following 115 different faiths.[] While Pennsylvania has a very large Amish population, Holmes County, Ohio has the largest Amish population in the world.[53] While Pennsylvania owes its existence to Quakers and many of the older trappings of the Commonwealth are rooted in the teachings of the Religious Society of Friends (as they are officially known), practicing Quakers are a small minority today.

The religious affiliations of the people of Pennsylvania:[]

- Christianity – 80%
  - Protestant – 51%
    - Mainline Protestant – 25%
    - Evangelical Protestant – 18%
    - Black Protestant – 7%
    - Jehovah Witness – 1%
  - Roman Catholic – 29%
  - Orthodox – 1%
- Non-religious/ Unaffiliated – 13%
- Judaism – 2%
- Other religions – 1%
- Don't know/ refused – 1%

According to a 2012 Gallup poll, 40% of Pennsylvanians are very religious, 28% are moderately religious, and 32% are non-religious.[54]

Economy

Pennsylvania's 2010 total gross state product (GSP) of $570 billion ranks the state 6th in the nation.[] If Pennsylvania were an independent country, its economy would rank as the 18th largest in the world. On a per-capita basis, Pennsylvania's per-capita GSP of $39,830 ranks 29th among the 50 states.[]

Bethlehem Steel's closed manufacturing facility in Bethlehem, Pennsylvania. This site later became a new multi-million dollar Sands Casino Resort in 2009 and became the site of the ArtsQuest Center at Steelstacks, Levitt Pavilion Steelstacks, and PBS 39 in 2011.

Philadelphia in the southeast corner, Pittsburgh in the southwest corner, Erie in the northwest corner, Scranton-Wilkes-Barre in the northeast corner, and Allentown-Bethlehem-Easton in the east central region are urban manufacturing centers. Much of the Commonwealth is rural; this dichotomy affects state politics as well as the state economy.[55] Philadelphia is home to six Fortune 500 companies,[] with more located in suburbs like King of Prussia; it's a leader in the financial[56] and insurance industry.

Pittsburgh is home to eight Fortune 500 companies, including U.S. Steel, PPG Industries, and H.J. Heinz.[] In all, Pennsylvania is home to fifty Fortune 500 companies.[] Erie is also home to GE Transportation Systems, which is the largest producer of train locomotives in the United States.

As in the US as a whole and in most states, the largest private employer in the Commonwealth is Wal-Mart, followed by the University of Pennsylvania.[57] [58]

As of April 2012, the state's unemployment rate is 7.4%.[59]

Banking

The first nationally chartered bank in the United States, the Bank of North America, was founded in 1781 in Philadelphia. After a series of mergers, the Bank of North America is part of Wells Fargo, which uses national charter 1.

Pennsylvania is also the home to the first nationally chartered bank under the 1863 National Banking Act. That year, the Pittsburgh Savings & Trust Company received a national charter and renamed itself the First National Bank of Pittsburgh as part of the National Banking Act. That bank is still in existence today as PNC Financial Services, and remains based in Pittsburgh. PNC is the state's largest bank, and the sixth-largest in the United States.

The famous Crayola crayon company is based in Easton.

Agriculture

Pennsylvania ranks 19th overall in agricultural production,[60] but 1st in mushrooms, 2nd in apples, 3rd in Christmas trees and layer chickens, 4th in nursery and sod, milk, corn for silage, grapes grown (including juice grapes),[] and horses production. It also ranks 8th in the nation in Winemaking.[]

Gambling

Casino gambling was legalized in Pennsylvania in 2004. Currently, there are nine casinos across the state with three under construction or in planning. Only horse racing, slot machines and electronic table games were legal in Pennsylvania, although a bill to legalize table games was being negotiated in the fall of 2009.[61] Tables games such as poker, roulette, black jack and dice were finally approved by the state legislature in January 2010, being signed into law by the Governor on January 7. Sports betting is illegal.

Governor Ed Rendell had considered legalizing video poker machines in bars and private clubs in 2009, since an estimated 17,000 operate illegally across the state.[62] Under this plan, any establishment with a liquor license would be allowed up to 5 machines. All machines would be connected to the state's computer system, like commercial casinos. The state would impose a 50% tax on net gambling revenues, after winning players have been paid, with the remaining 50% going to the establishment owners.

Film

The Pennsylvania Film Production Tax Credit began in 2004 and stimulated the development of a film industry in the state.

Governance

Presidential elections results[63]

Year	Republican	Democratic
2012	46.58% *2,680,434*	**51.97%** *2,990,274*
2008	44.15% *2,655,885*	**54.47%** *3,276,363*
2004	48.42% *2,793,847*	**50.92%** *2,938,095*
2000	46.43% *2,281,127*	**50.60%** *2,485,967*
1996	39.97% *1,801,169*	**49.17%** *2,215,819*
1992	36.13% *1,791,841*	**45.15%** *2,239,164*
1988	**50.70%** *2,300,087*	48.39% *2,194,944*
1984	**53.34%** *2,584,323*	45.99% *2,228,131*
1980	**49.59%** *2,261,872*	42.48% *1,937,540*
1976	47.73% *2,205,604*	**50.40%** *2,328,677*
1972	**59.11%** *2,714,521*	39.13% *1,796,951*
1968	44.02% *2,090,017*	**47.59%** *2,259,405*
1964	34.70% *1,673,657*	**64.92%** *3,130,954*
1960	48.74% *2,439,956*	**51.06%** *2,556,282*

Voter Registration and Party Enrollment as of June 28, 2010[64]			
Party		Number of Voters	Percentage
	Democratic	4,309,604	51.00%
	Republican	3,122,036	36.95%
	Unaffiliated	492,077	5.82%
	Other Parties	525,962	6.22%
Total		**8,449,679**	**100%**

Government

Pennsylvania has had five constitutions during its statehood:[] 1776, 1790, 1838, 1874, and 1968 [65]. Prior to that, the province of Pennsylvania was governed for a century by a Frame of Government, of which there were four versions: 1682, 1683, 1696, and 1701.[] The capital of Pennsylvania is Harrisburg. The legislature meets in the State Capitol there.

Since 1992, Pennsylvania has been trending Democratic in Presidential elections (though the Pittsburgh metropolitan area trended more Republican in the 2008 Presidential election), voting for Bill Clinton twice by large margins, and slightly closer in 2000 for Al Gore. In the 2004 Presidential Election, Senator John F. Kerry beat President George W. Bush in Pennsylvania 2,938,095 (50.92%) to 2,793,847 (48.42%). Most recently, in the 2008 Presidential Election, Democrat Barack Obama defeated Republican John McCain in Pennsylvania, 3,184,778 (54%) to 2,584,088 (44%). The state holds 20 electoral votes.[]

In recent elections, Pennsylvania has leaned Democratic. The state has voted for the Democratic ticket for president in every election since 1992 and during the 2008 election campaign a recruitment drive saw registered Democrats outnumber registered Republicans by 1.2 million. However, Pennsylvania has a history of electing Republican senators. From 2009 to 2011, the state was represented by two Democratic senators for the first time since 1947. In 2010, Republicans recaptured a U.S. Senate seat as well as a majority of the state's congressional seats, control of both chambers of the state legislature and the governor's mansion. Democratic political consultant James Carville once pejoratively described Pennsylvania as "Philadelphia in the east, Pittsburgh in the west and Alabama in the middle".

Governor

The current Governor is Tom Corbett, former Attorney General of Pennsylvania. The other elected officials composing the executive branch are the Lieutenant Governor Jim Cawley, Attorney General Kathleen Kane, Auditor General Eugene DePasquale, and State Treasurer Robert McCord.[66]

General Assembly

Pennsylvania has a bicameral legislature set up by Commonwealth's constitution in 1790. The original Frame of Government of William Penn had a unicameral legislature.[67] The General Assembly includes 50 Senators[68] and 203 Representatives.[69] Joe Scarnati is currently President Pro Tempore of the State Senate,[70] Dominic Pileggi the Majority Leader,[71] and Jay Costa the Minority Leader.[72] Sam Smith is Speaker of the House of Representatives,[73] with Mike Turzai as Majority Leader[74] and Frank Dermody as Minority Leader.[75] As of the 2012 elections, the Republicans hold the majority in the State House and Senate.

Judiciary

Pennsylvania is divided into 60 judicial districts,[] most of which (except Philadelphia) have magisterial district judges (formerly called district justices and justices of the peace), who preside mainly over preliminary hearings in felony and misdemeanor offenses, all minor (summary) criminal offenses, and small civil claims.[] Most criminal and civil cases originate in the Courts of Common Pleas, which also serve as appellate courts to the district judges and for local agency decisions.[] The Superior Court hears all appeals from the Courts of Common Pleas not expressly designated to the Commonwealth Court or Supreme Court. It also has original jurisdiction to review warrants for wiretap surveillance.[] The Commonwealth Court is limited to appeals from final orders of certain state agencies and certain designated cases from the Courts of Common Pleas.[] The Supreme Court of Pennsylvania is the final appellate court. All judges in Pennsylvania are elected; the chief justice is determined by seniority.[]

Taxation

Pennsylvania has the 10th highest tax burden in the United States.[76] Residents pay a total of $83.7 billion in state and local taxes with a per capita average of $6,640 annually. Residents share 76% of the total tax burden. Many state politicians have tried to increase the share of taxes paid by out of state sources. Suggested revenue sources include taxing natural gas drilling as Pennsylvania is the only state without such a taxation on gas drilling.[77] Additional revenue prospects include trying to place tolls on interstate highways; specifically Interstate 80 which is used heavily by out of state commuters with high maintenance costs.[78]

Sales tax provides 39% of Commonwealth's revenue; personal income tax 34%; motor vehicle taxes about 12%, and taxes on cigarettes and alcohol beverage 5%.[79] Personal income tax is a flat 3.07%. An individual's taxable income is based on the following eight types of income: compensation (salary); interest; dividends; net profits from the operation of a business, profession or farm; net gains or income from the dispositions of property; net gains or income from rents, royalties, patents and copyrights; income derived through estates or trusts; and gambling and lottery winnings (other than Pennsylvania Lottery winnings).[80]

Counties, municipalities, and school districts levy taxes on real estate. In addition, some local bodies assess a wage tax on personal income. Generally, the total wage tax rate is capped at 1% of income but some municipalities with home rule charters may charge more than 1%. Thirty-two of the Commonwealth's sixty-seven counties levy a personal property tax on stocks, bonds, and similar holdings.

Representation in the 113th Congress

Pennsylvania's two U.S. Senators in the 113th Congress are Bob Casey, Jr. and Pat Toomey.

Pennsylvania's U.S. Representatives for the term beginning January 2013 are Bob Brady (1st), Chaka Fattah (2nd), Mike Kelly (3rd), Scott Perry (politician) (4th), Glenn "G.T." Thompson (5th), Jim Gerlach (6th), Pat Meehan (7th), Mike Fitzpatrick (8th), Bill Shuster (9th), Tom Marino (10th), Lou Barletta (11th), Keith Rothfus (12th), Allyson Schwartz (13th), Mike Doyle (14th), Charlie Dent (15th), Joe Pitts (16th), Matt Cartwright (17th), Tim Murphy (18th).[81]

See map of congressional districts

Municipalities

Pennsylvania is divided into 67 counties.[82] Counties are further subdivided into municipalities that are either incorporated as cities, boroughs, or townships.[83] One county, Philadelphia County, is coterminous with the city of Philadelphia after it was consolidated in 1854.

The skyline of Philadelphia, the largest city in Pennsylvania.

There are a total of 56 cities in Pennsylvania, which are classified, by population, as either first, second, or third class cities.[82] [84] Philadelphia, Pennsylvania's largest city, has a population of 1,547,297 and is the state's only first class city.[83] Pittsburgh (305,647) and Scranton (76,072) are second class and second class 'A' cities, respectively.[83]

The skyline of Pittsburgh, the second largest city in Pennsylvania.

The rest of the cities, like the third and fourth largest—Allentown (107,815) and Erie (103,571)—to the smallest—Parker with a population of only 738—are third class cities.[85] First and second class cities are governed by a "strong mayor" form of mayor–council government, whereas third class cities are governed by either a "weak mayor" form of government or a council–manager government.[83]

Boroughs are generally smaller than cities, with most Pennsylvania cities having been incorporated as a borough before being incorporated as a city.[83] There are 958 boroughs in Pennsylvania, all of which governed by the "weak mayor" form of mayor–council government.[82] [83]

The skyline of Allentown, the third largest city in Pennsylvania.

Townships are the third type of municipality in Pennsylvania and are classified as either first class or second class townships. There are 1,454 second class townships and 93 first class townships.[86] Second class township can become first class townships if it has a population density greater than 300 inhabitants per square mile (/km^2) and a referendum is passed supporting the change.[86]

There is one exception to the types of municipalities in Pennsylvania: Bloomsburg was incorporated as a town in 1870 and is, officially, the only town in the state.[87] In 1975, McCandless Township adopted a home-rule charter under the name of "Town of McCandless", but is, legally, still a first class township.[88]

The total of 56 cities, 958 boroughs, 93 first class townships, 1454 second class townships, and 1 town (Bloomsburg) is 2562 municipalities.

Education

Pennsylvania has 500 public school districts, thousands of private schools, publicly funded colleges and universities, and over 100 private institutions of higher education.

The University of Pennsylvania quadrangle in Philadelphia

Primary and secondary education

In general, under state law, school attendance in Pennsylvania is mandatory for a child from the age of 8 until the age of 17, or until graduation from an accredited high school, whichever is earlier.[89] As of 2005, 83.8% of Pennsylvania residents age 18 to 24 have completed high school. Among residents age 25 and over, 86.7% have graduated from high school. Additionally, 25.7% have gone on to obtain a bachelor's degree or higher.[90] State students consistently do well in standardized testing. In 2007, Pennsylvania ranked 14th in mathematics, 12th in reading, and 10th in writing for 8th grade students.[91]

In 1988, the Pennsylvania General Assembly passed Act 169, which allows parents or guardians to homeschool their children as an option for compulsory school attendance. This law specifies the requirements and responsibilities of the parents and the school district where the family lives.[92]

Higher education

The Pennsylvania State System of Higher Education (PASSHE) is the public university system of the Commonwealth, with 14 state-owned schools. The Commonwealth System of Higher Education is organizing body of the 4 state-related schools in Pennsylvania, these schools are independent institutions that receive some state funding. There are also 15 publicly funded two-year community colleges and technical schools that are separate from the PASSHE system. Additionally there are many private two- and four-year technical schools, colleges and universities.

Carnegie Mellon University, The Pennsylvania State University, the University of Pennsylvania, and the University of Pittsburgh, are members of the Association of American Universities, an invitation only organization of leading research universities. The Pennsylvania State University is the Commonwealth's Land-grant university, Sea Grant College and, Space Grant College. The University of Pennsylvania, located in Philadelphia, is considered the first university in the United States and established the country's first medical school. The University of Pennsylvania is also the Commonwealth's only, and geographically the most southern, Ivy League school. The Pennsylvania Academy of the Fine Arts is the first and oldest art school in the United States.[] Philadelphia College of Pharmacy, now a part of University of the Sciences in Philadelphia, was the first pharmacy school in the United States.[]

Recreation

Pennsylvania is home to the nation's first zoo, the Philadelphia Zoo.[93] Other long-accredited AZA zoos include the Erie Zoo and the Pittsburgh Zoo & PPG Aquarium. The Lehigh Valley Zoo and ZOOAMERICA are other notable zoos. The Commonwealth boasts some of the finest museums in the country, including the Carnegie Museums in Pittsburgh, the Philadelphia Museum of Art, and several others. One unique museum is the Houdini Museum in Scranton, the only building in the world devoted to the legendary magician.[94] Pennsylvania is also home to the National Aviary, located in Pittsburgh.

Dorney Park and Wildwater Kingdom's Steel Force and Thunderhawk roller coasters in Allentown

All 121 state parks in Pennsylvania feature free admission.

Pennsylvania offers a number of notable amusement parks, including Camel Beach, Conneaut Lake Park, Dorney Park & Wildwater Kingdom, Dutch Wonderland, DelGrosso Amusement Park, Hersheypark, Idlewild Park, Kennywood, Knoebels, Lakemont Park, Sandcastle Waterpark, Sesame Place, Great Wolf Lodge and Waldameer Park. Pennsylvania also is home to the largest indoor waterpark resort on the East Coast, Splash Lagoon in Erie.

Ohiopyle State Park timelapse

There are also notable music festivals that take place in Pennsylvania. These include Musikfest and NEARfest in Bethlehem, the Philadelphia Folk Festival, Creation Festival, the Great Allentown Fair, and Purple Door.

There are nearly one million licensed hunters in Pennsylvania. Whitetail deer, cottontail rabbits, squirrel, turkey, and grouse are common game species. Pennsylvania is considered one of the finest wild turkey hunting states in the Union, alongside Texas and Alabama. Sport hunting in Pennsylvania provides a massive boost for the Commonwealth's economy. A report from The Center for Rural Pennsylvania (a Legislative Agency of the Pennsylvania General Assembly) reported that hunting, fishing, and furtaking generated a total of $9.6 billion statewide.

The Boone and Crockett Club shows that five of the ten largest (skull size) black bear entries came from the state.[] The state also has a tied record for the largest hunter shot black bear in the Boone & Crockett books at 733 lb (kg) and a skull of 23 3/16 tied with a bear shot in California in 1993.[] The largest bear ever found dead was in Utah in 1975, and the second largest was shot by a poacher in the state in 1987.[] Pennsylvania holds the second highest number of Boone & Crockett-recorded record black bears at 183, second only to Wisconsin's 299.[]

Transportation

Exterior

Interior

The Pennsylvania Department of Transportation, abbreviated as PennDOT, owns 39861 miles (km) of the 121770 miles (km) of roadway in the state, making it the fifth largest state highway system in the United States.[95] The Pennsylvania Turnpike system is 535 miles (km) long, with the mainline portion stretching from Ohio to Philadelphia and New Jersey.[95] It is overseen by the Pennsylvania Turnpike Commission. Another major east–west route is Interstate 80, whichs runs primarily in the northern tier of the state from Ohio to New Jersey at the Delaware Water Gap. Interstate 90 travels the relatively short distance between Ohio and New York through Erie County, in the extreme northwestern part of the state.

U.S. Route 220 as it passes through Lamar Township.

Primary north–south highways are Interstate 79 from its terminus in Erie through Pittsburgh to West Virginia, Interstate 81 from New York through Scranton, Lackawanna County and Harrisburg to Maryland and Interstate 476, which begins 7 miles (km) north of the Delaware border, in Chester, Delaware County and travels 132 miles (km) to Clarks Summit, Lackawanna County, where it joins I-81. All but 20 miles (km) of I-476 is the Northeast Extension of the Pennsylvania Turnpike, while the highway south of the main line of the Pennsylvania Turnpike is officially called the "Veterans Memorial Highway", but is commonly referred to by locals as the "Blue Route".

The Southeastern Pennsylvania Transportation Authority (SEPTA) is the sixth largest transit agency in the United States and operates the commuter, heavy and light rail transit, and transit bus service in the Philadelphia metropolitan area. The Port Authority of Allegheny County is the 25th largest transit agency and provides transit bus and light rail service in and around Pittsburgh.[96]

Intercity passenger rail transit is provided by Amtrak, with the majority of traffic occurring on the *Keystone Service* in the high-speed Keystone Corridor between Harrisburg and Philadelphia's 30th Street Station before heading north to New York City; the *Pennsylvanian* follows the same route from New York City to Harrisburg, but extends out to Pittsburgh. The *Capitol Limited* also passes through Pittsburgh, as well as Connellsville, on its way from Chicago to Washington, D.C.[97] Traveling between Chicago and New York City, the *Lake Shore Limited* passes through Erie once in each direction.[97] There are 67 short-line, freight railroads operating in Pennsylvania, the highest number in any U.S. state.[97]

Pennsylvania has seven major airports: Philadelphia International, Pittsburgh International, Lehigh Valley International, Harrisburg International, Erie International, University Park Airport and Wilkes-Barre/Scranton International. A total of 134 public-use airport are located in the state.[97] The port of Pittsburgh is the second largest inland port in the United States and the 18th largest port overall; the Port of Philadelphia is the 24th largest port in the United States.[98] Pennsylvania's only port on the Great Lakes is located in Erie.

The Allegheny River Lock and Dam Two is the most-used lock operated by the United States Army Corps of Engineers of its 255 nationwide.[99] The dam impounds the Allegheny River near Downtown Pittsburgh.

Sports

Further information: List of people from Pennsylvania

Pennsylvania is home to many professional sports teams; the Pittsburgh Steelers and Philadelphia Eagles of the National Football League, the Philadelphia Phillies and Pittsburgh Pirates of Major League Baseball, the Philadelphia 76ers of the National Basketball Association, the Philadelphia Flyers and Pittsburgh Penguins of the National Hockey League, the Philadelphia Union of Major League Soccer, the Erie Bayhawks of the National Basketball Association Development League, the Wilkes-Barre/Scranton Penguins and Hershey Bears of the American Hockey League, the Philadelphia Soul and the Pittsburgh Power of the Arena Football League, and the Philadelphia Wings of the National Lacrosse League. Among them, these teams have accumulated 7 World Series Championships (Pirates 5, Phillies 2), 16 National League Pennants (Pirates 9, Phillies 7), 3 pre-Super Bowl era NFL Championships (Eagles), 6 Super Bowl Championships (Steelers), 1 Arena Bowl Championship (Soul), 2 NBA Championships (76ers), 5 Stanley Cups (Penguins 3, Flyers 2), 11 Calder Cups (Bears), and 6 Champion's Cups (Wings).

There are many minor league baseball teams located throughout the state; several of these teams are associated with either the Phillies or the Pirates. In 2008, the Phillies moved their AAA-level team, the Lehigh Valley IronPigs, from Ottawa, Ontario, in Canada, to a newly constructed stadium, Coca-Cola Park in Allentown. The Lehigh Valley is a core fan base for both the Phillies and the Philadelphia Eagles, who conduct their pre-season training camp on the practice fields of Lehigh University. Therefore, expectations are that the Lehigh Valley IronPigs (named after pig iron, an instrumental part in the construction of steel which used to be a large part of the local economy for decades), is likely to prove popular among Allentown and Lehigh Valley Phillies fans.

The Phillies' AA team, also called the Phillies, is located in Reading, while the short-season A-level affiliate, called the Crosscutters, is located in Williamsport. The Pirates' AA team, the Curve, is located in Altoona. The short-season A-level affiliate, the State College Spikes, is located in State College. The Spikes share a stadium with the Penn State University baseball team. Other Major League Baseball teams have a presence in the state as well. The New York Yankees' AAA team, also called the Yankees, is located in Moosic, between Scranton and Wilkes-Barre in the northeastern part of the state. The Detroit Tigers' AA team, the SeaWolves, is located in Erie, and the Washington Nationals' AA affiliate, the Senators, plays in the capital of Harrisburg. Two independent-league teams, the Lancaster Barnstormers and York Revolution of the Atlantic League of Professional Baseball, are located in south-central Pennsylvania, while the Washington Wild Things of the Frontier League are located in the south-western corner of the state.

Coca-Cola Park in Allentown, home of the Lehigh Valley IronPigs

Each summer, the Little League World Series is held in South Williamsport, near where Little League Baseball was founded in Williamsport. Also, the first World Series between the Boston Pilgrims (which became the Boston Red Sox) and Pittsburgh Pirates was played in Pittsburgh in 1903.

College football is very popular in Pennsylvania. The Penn State University Nittany Lions were coached by Joe Paterno who has led Penn State to two national championships (1982 & 1986) as well as five undefeated seasons (1968, 1969, 1973, 1986 and 1994). On January 22, 2012 Joe Paterno passed away from lung cancer. Penn State now has a new coach, Bill O'Brien, who was serving as the offensive coordinator for the New England Patriots before becoming head coach of Penn State. Penn State plays its home games in the second largest stadium in the United States, Beaver Stadium, which seats 107,282. In addition, the University of Pittsburgh Panthers have won nine national championships (1915, 1916, 1918, 1929, 1931, 1934, 1936, 1937 and 1976) and have played eight undefeated seasons (1904, 1910, 1915, 1916, 1917, 1920, 1937 and 1976).[100] Pitt plays its home games at Heinz Field, a facility it shares with the Pittsburgh Steelers. Other Pennsylvania schools that have won national titles in football include Lafayette College (1896), Villanova University (FCS 2009), and the University of Pennsylvania (1895, 1897, 1904 and 1908).[101]

College basketball is also popular in the state, especially in the Philadelphia area where five universities, collectively termed the Big Five, have a rich tradition in NCAA Division I basketball. National titles in college basketball have been won by the following Pennsylvania universities: La Salle University (1954), Temple University (1938), University of Pennsylvania (1920 and 1921), University of Pittsburgh (1928 and 1930) and Villanova University (1985).[102] [103]

Soccer is gaining popularity within the state of Pennsylvania as well. With the addition of the Philadelphia Union in the MLS, the state now boasts three teams that are eligible to compete for the Lamar Hunt U.S. Open Cup annually. The other two teams are the Pittsburgh Riverhounds and the Harrisburg City Islanders, both of the United Soccer Leagues Second Division (USL-2). Within the American Soccer Pyramid, the MLS takes the first tier, while the USL-2 claims the third tier.

In motorsports, the Mario Andretti dynasty of race drivers hails from Nazareth in the Lehigh Valley. Notable Racetracks in Pennsylvania include the Jennerstown Speedway in Jennerstown, the Lake Erie Speedway in North East, the Mahoning Valley Speedway in Lehighton, the Motordome Speedway in Smithton, the Mountain Speedway in St. Johns, the Nazareth Speedway in Nazareth; and the Pocono Raceway in Long Pond, which is home to both the NASCAR sanctioned Pennsylvania 500 and Pocono 500 stock car races. The state is also home to Maple Grove Raceway, near Reading, which hosts major National Hot Rod Association sanctioned drag racing events each year.

The Pocono 500 at Pocono Raceway in Long Pond

There are also two motocross race tracks that host a round of the AMA Toyota Motocross Championships in Pennsylvania. High Point Raceway [104] in located in Mt. Morris, Pennsylvania, and Steel City is located in Delmont, Pennsylvania.

Horse racing courses in Pennsylvania consist of The Meadows near Pittsburgh, Pocono Downs in Wilkes-Barre, and Harrah's Philadelphia in Chester, which offer harness racing, and Penn National Race Course in Grantville, Parx Racing (formerly Philadelphia Park) in Bensalem, and Presque Isle Downs near Erie, which offer thoroughbred racing. Smarty Jones, the 2004 Kentucky Derby and Preakness Stakes winner, had Philadelphia Park as his home course.

Arnold Palmer, one of the 20th century's most notable pro golfers, comes from Latrobe, while Jim Furyk, a current PGA member, grew up near in Lancaster. PGA tournaments in Pennsylvania include the 84 Lumber Classic, played at Nemacolin Woodlands Resort, in Farmington and the Northeast Pennsylvania Classic, played at Glenmaura National Golf Club, in Moosic.

Philadelphia is home to LOVE Park, once a skateboarding mecca, and across from City Hall, host to ESPN's X Games in 2001 and 2002.[105]

Food

In his book *Yo Mama Cooks Like a Yankee*, author Sharon Hernes Silverman calls Pennsylvania the snack food capital of the world.[106] It leads all other states in the manufacture of pretzels and potato chips. The Sturgis Pretzel House introduced the pretzel to America, and companies like Anderson Bakery Company, Intercourse Pretzel Factory, and Snyder's of Hanover are leading manufacturers in the Commonwealth. Two of the three companies that define the U.S. potato chip industry are based in Pennsylvania: Utz Quality Foods, Inc., which started making chips in Hanover, Pennsylvania in 1921, and Wise Snack Foods which started making chips in Berwick in 1921 (the third, Lay's Potato Chips, is a Texas company). Other companies such as Herr Foods, Martin's Potato Chips, Snyder's of Berlin (not associated with Snyder's of Hanover) and Troyer Farms Potato Products are popular chip manufacturers.

The U.S. chocolate industry is centered in Hershey, Pennsylvania, with Mars, Godiva, and Wilbur Chocolate Company nearby, and smaller manufacturers such as Asher's[107] in Souderton,[108] and Gertrude Hawk of Dunmore. Other notable companies include Just Born in Bethlehem, Pennsylvania, makers of Hot Tamales, Mike and Ikes, the Easter favorite marshmallow Peeps, and Boyer Brothers of Altoona, Pennsylvania, which is well known for its Mallo Cups. Auntie Anne's Pretzels began as a market-stand in Downingtown, Pennsylvania, and now has corporate headquarters in Lancaster City.[109] Traditional Pennsylvania Dutch foods include chicken potpie, ham potpie, schnitz un knepp (dried apples, ham, and dumplings), fasnachts (raised doughnuts), scrapple, pretzels, bologna, chow-chow, and Shoofly pie. Martin's Famous Pastry Shoppe, Inc., headquartered in Chambersburg, Pennsylvania, specializes in potato bread, another traditional Pennsylvania Dutch food. D.G. Yuengling & Son, America's oldest brewery, has been brewing beer in Pottsville since 1829.

Among the regional foods associated with Philadelphia are cheesesteaks, hoagie, soft pretzels, Italian water ice, scrapple, Tastykake, and strombolis. In Pittsburgh, tomato ketchup was improved by Henry John Heinz from 1876 to the early 20th century. Famous to a lesser extent than Heinz ketchup are the Pittsburgh's Primanti Brothers Restaurant sandwiches, pierogies, and city chicken. Outside of Scranton, in Old Forge there are dozens of Italian restaurants specializing in pizza made unique by thick, light crust and American cheese. Erie also has its share of unique foods, including Greek sauce and sponge candy. Sauerkraut along with pork and mashed potatoes is a common meal on New Year's Day in Pennsylvania.

State symbols

The Ruffed Grouse

- **Motto**: "Virtue, liberty, and independence"
- **Tree**: Eastern Hemlock[110]
- **Game bird**: Ruffed Grouse[111]
- **Flower**: Mountain Laurel[111]
- **Insect**: Pennsylvania Firefly[111]
- **Animal**: White-tailed deer[111]
- **Dog**: Great Dane[111]
- **Fish**: Brook Trout[111]
- **Fossil**: *Phacops rana*[110]
- **Beverage**: Milk[110]
- **Song**: "Pennsylvania"[112]
- **Ship**: US Brig *Niagara*[110]
- **Electric locomotive**: GG1 4859[110]
- **Steam locomotive**: K4s 1361 and K4s 3750[110]
- **Beautification plant**: Penngift Crownvetch[110]

Nicknames

US Brig *Niagara* in port

Pennsylvania has been known as the *Keystone State* since 1802,[113] based in part upon its central location among the original Thirteen Colonies forming the United States, and also in part because of the number of important American documents signed in the state (such as the Declaration of Independence). It was also a keystone state economically, having both the industry common to the North (making such wares as Conestoga wagons and rifles)[114] [115] and the agriculture common to the South (producing feed, fiber, food, and tobacco).[116]

Another one of Pennsylvania's nicknames is the *Quaker State*; in colonial times, it was known officially as the *Quaker Province*,[117] in recognition of Quaker[118] William Penn's *First Frame of Government*[119] constitution for Pennsylvania that guaranteed liberty of conscience. He knew of the hostility[120] Quakers faced when they opposed religious ritual, taking oaths, violence, war and military service, and what they viewed as ostentatious frippery.[121]

"**The Coal State**", "**The Oil State**", "**The Chocolate State**", and "**The Steel State**" were adopted when those were the state's greatest industries.[122]

"The State of Independence" currently appears on many road signs entering the state.

Gallery

Aliquippa

Allentown

Altoona

Beaver Falls

Bedford

Bethlehem

Blairsville

Butler

Carlisle

Chambersburg

Clarion

Clearfield

Corry

Coudersport

Easton

Ebensburg

Edinboro

Erie

Franklin

Gettysburg

Greencastle

Greensburg

Hanover

Hazleton

Hershey

Hollidaysburg

Indiana

Jeannette

Johnstown

Kittanning

Lancaster

Latrobe

Lebanon

Lewisburg

Lewistown

Lock Haven

McKees Rocks

Meadville

Monessen

New Castle

New Hope

Palmyra

Philadelphia

Phoenixville

Pittsburgh

Pittston City

Pottstown

Pottsville

Punxsutawney

Reading

Renovo

Ridgway

Scranton

Selinsgrove

Shamokin

State College

Sunbury

Tyrone

Uniontown

Warren

Washington

Waynesboro

Wilkes-Barre

Williamsport

York

See also

- Outline of Pennsylvania – organized list of topics about Pennsylvania
- Index of Pennsylvania-related articles

Notes

[1] 2006 Statistical Abstract: Geography & Environment: Land and Land Use (http://www.census.gov/compendia/statab/tables/06s0347.xls)

[3] http://www.nws.noaa.gov/climate/xmacis.php?wfo=phi

[4] http://www.nws.noaa.gov/climate/xmacis.php?wfo=bgm

[5] http://www.nws.noaa.gov/climate/xmacis.php?wfo=ctp

[6] http://www.nws.noaa.gov/climate/xmacis.php?wfo=pbz

[7] http://www.nws.noaa.gov/climate/xmacis.php?wfo=cle

[25] Charter for the Province of Pennsylvania-1681 (http://avalon.law.yale.edu/17th_century/pa01.asp). This charter, granted by Charles II (England) to William Penn, constituted him and his heirs proprietors of the province, which, in honor of his father, Admiral William Penn, (whose cash advances and services were thus requited,) was called Pennsylvania. To perfect his title, William Penn purchased, on 1682-08-24, a quit-claim from the Duke of York to the lands west of the Delaware River embraced in his patent of 1664

[31] David Yount (2007). " *How the Quakers invented America* (http://books.google.com/books?id=pk7ycUq3cxsC&pg=PA82&dq&hl=en#v=onepage&q=&f=false)". Rowman & Littlefield. p.82. ISBN 0-7425-5833-9

[32] Sydney G. Fisher (2009). " *The Quaker Colonies* (http://books.google.com/books?id=zKzFgAlx1CkC&pg=PA13&dq&hl=en#v=onepage&q=&f=false)". Echo Library. p.13 ISBN 1-4068-5110-8

[33] Hamilton, Alexander and Syrett, Harold C. *The Papers of Alexander Hamilton*. 1963, page 240.

[41] Battle of Gettysburg (http://content.answers.com/main/content/wp/en/thumb/b/b3/200px-Dddr66.jpg)

[42] American FactFinder – Results (http://factfinder2.census.gov/bkmk/table/1.0/en/ACS/10_SF4/DP02/0400000US42)

[44] " Americans under age 1 now mostly minorities, but not in Ohio: Statistical Snapshot (http://www.cleveland.com/datacentral/index.ssf/2012/06/americas_under_age_1_populatio.html)". *The Plain Dealer*. June 3, 2012.

[48] http://www.census.gov/popest/states/asrh/tables/SC-EST2005-03-42.csv
[51] The Works of Voltaire, volume 19 (http://oll.libertyfund.org/Texts/Voltaire0265/Works/Vol19-HTMLs/0060-19_Pt02.html)
[52] Religious diversity in Pennsylvania (http://cip.cornell.edu/Dienst/UI/1.0/Summarize/psu.ph/1134140590)
[55] Appeals court races wrap up with focus on voter mobilization (http://www.phillyburbs.com/pb-dyn/news/103-11042003-190258.html)
[59] (http://www.deptofnumbers.com/unemployment/pennsylvania/)| title= Department of Numbers | accessdate=May 27, 2012
[60] Agricultural Census 2002 (http://www.nass.usda.gov/census/census02/profiles/pa/cp99042.PDF)
[61] (http://www.kyw1060.com/Pa--Lawmakers-Consider-Table-Games--More-Small-Cas/5763077)
[65] http://sites.state.pa.us/PA_Constitution.html
[66] Commonwealth of PA - Election Results 2012 (http://www.electionreturns.state.pa.us/ElectionsInformation.aspx?FunctionID=12&ElectionID=53)
[79] *Revenue Department Releases August Collections (09/01/2006)* (http://www.revenue.state.pa.us/revenue/CWP/view.asp?Q=261929&A=208). Retrieved September 25, 2006.
[82] *The Pennsylvania Manual*, p. 6-3.
[83] *Pennsylvania Manual*, p. 6-5.
[84] *The Pennsylvania Manual*, p. 6-46.
[86] *The Pennsylvania Manual*, p. 6-6.
[87] *The Pennsylvania Manual*, p. 6-22.
[88] Title 302, Pennsylvania Code, Section 23.1–101.
[89] (http://www.pde.state.pa.us/) 'Pennsylvania Department of Education (PDE). Retrieved on December 4, 2009.'
[90] (http://nces.ed.gov/) 'National Center for Education Statistics (NCES). Retrieved on December 4, 2009.'
[91] 'NCES.'
[92] (http://www.pde.state.pa.us/home_education/site/default.asp) 'Pennsylvania Department of Education: Home Education and Private Tutoring. Retrieved on December 4, 2009.'
[93] (http://www.philadelphiazoo.org/index.php?id=10_2_1)
[95] "Pennsylvania Department of Transportation Fact Book", p. 7.
[96] "2010 Public Transportation Fact Book", p. 8.
[97] "Pennsylvania Department of Transportation Fact Book", p. 10.
[98] Waterborne Commerce Statistics Center, p. 5-4.
[104] http://www.highpointmx.com/
[107] http://news.google.com/newspapers?id=hsQfAAAAIBAJ&sjid=i9gEAAAAIBAJ&pg=1978,4428241&dq=asher%27s-chocolates&hl=en
[110] *The Pennsylvania Manual*, p. 1-6
[111] *The Pennsylvania Manual*, p. 1-5
[112] *The Pennsylvania Manual*, p. 1-7
[113] PHMC: State Symbols (http://www.phmc.state.pa.us/bah/pahist/symbols.asp?secid=31)
[115] Lancaster rifle (http://www.2nj.org/library/weapons/flintlock_rifles.htm)
[116] PHMC: Agriculture in Pennsylvania (http://www.phmc.state.pa.us/ppet/agriculture/page1.asp?secid=31)
[117] The Quaker Province (http://www.dgs.state.pa.us/dgs/lib/dgs/pa_manual/section1/the_quaker_province_1681-1776.pdf#search="quaker province ")
[120] Pennsylvania translates to "Penn's Woods" and was named after the father of William Penn, the founder of the colony. Digital History: Persecution of the Quakers (http://www.digitalhistory.uh.edu/documents/documents_p2.cfm?doc=240)
[121] The Quaker Province 1681–1776 (http://www.phmc.state.pa.us/bah/pahist/quaker.asp?secid=31)

External links

- Pennsylvania (http://www.dmoz.org/Regional/North_America/United_States/Pennsylvania/) at the Open Directory Project
- Gov. Andrew Curtin's Pennsylvania Reserve Volunteer Corps, Civil War 1861–1864 (http://www.pareserves.com/)
- Official state government site (http://www.pa.gov/)
- Pennsylvania Department of Transportation (http://www.dot.state.pa.us/)
- Allegheny National Forest (http://www.fs.fed.us/r9/forests/allegheny/)
- Pennsylvania Wilds (http://www.pawilds.com/)
- USGS real-time, geographic, and other scientific resources of Pennsylvania (http://www.usgs.gov/state/state.asp?State=PA)
- Energy Data & Statistics for Pennsylvania (http://tonto.eia.doe.gov/state/state_energy_profiles.cfm?sid=PA)

- Pennsylvania State Facts (http://www.ers.usda.gov/StateFacts/PA.htm)
- Official state tourism site (http://www.visitPA.com/)
- Biography of William Penn from 1829 (http://www.antiquebooks.net/readpage.html#penn)
- A History of Pennsylvania from 1905 (http://rack1.ul.cs.cmu.edu/is/pennhist/)
- Free Original Documents Online: Pennsylvania State Archives 1600s to 1800s (http://www.footnote.com/topicpage.php?tp=88)
- Pennsylvania Department of Community and Economic Development (http://newpa.com/)
- National Association of Counties (information on each Pennsylvania County)

Population_density

Population density (in agriculture **standing stock** and standing crop) is a measurement of population per unit area or unit volume. It is frequently applied to living organisms, and particularly to humans. It is a key geographic term.[1]

Population density (people per km^2) by country, 2006

Biological population densities

Population density is population divided by total land area or water volume, as appropriate.[1]

Low densities may cause an extinction vortex and lead to further reduced fertility. This is called the Allee effect after the scientist who identified it. Examples of the causes in low population densities include:[2]

- Increased problems with locating sexual mates
- Increased inbreeding

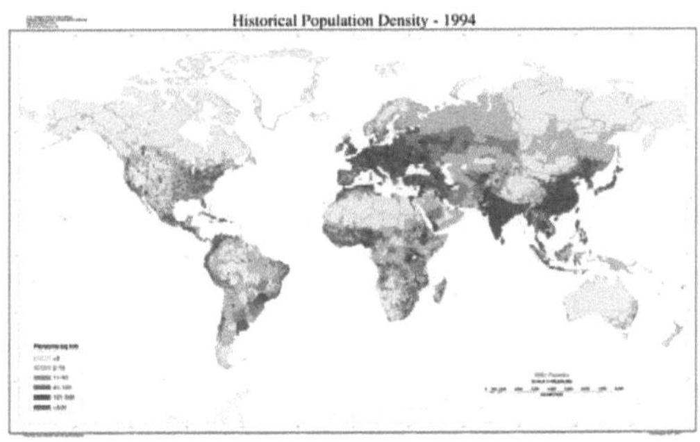

Population density (people per km^2) map of the world in 1994 (detailed).

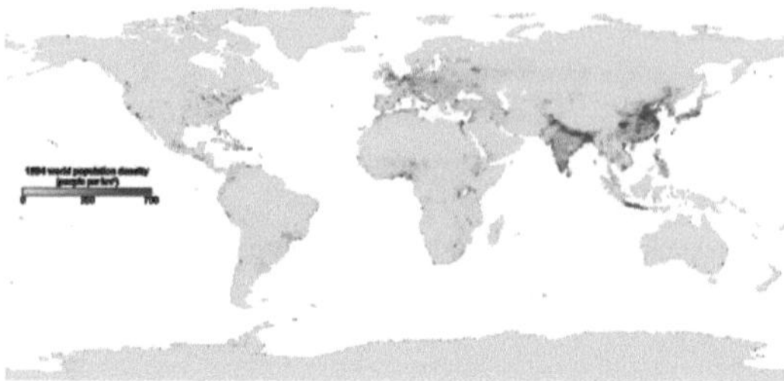

Population density (people per km^2) map of the world in 1994.

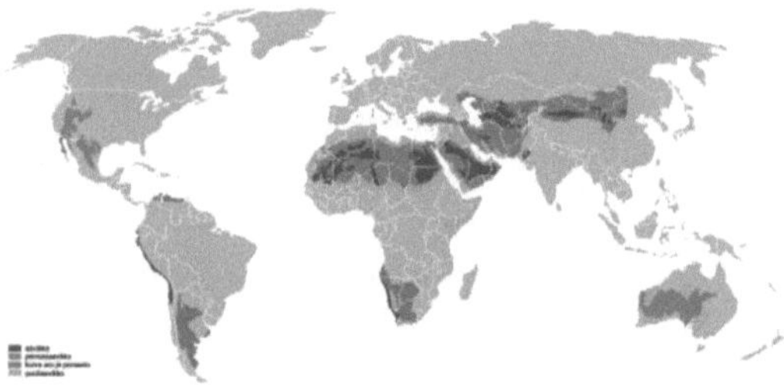

Deserts around the world. Compare with maps above.

Human population density

Monaco in South Europe, currently holds the record for being the most densely populated nation in the world.

For humans, population density is the number of people per unit of area usually per square kilometer or mile (which may include or exclude cultivated or potentially productive area). Commonly this may be calculated for a county, city, country, another territory, or the entire world.

The world's population is 7 billion,[3] and Earth's total area (including land and water) is 510 million square kilometers (197 million square miles).[4] Therefore the worldwide human population density is 6.8 billion ÷ 510 million = 13.3 per km^2 (34.5 per sq. mile). If only the Earth's land area of 150 million km^2 (58 million sq. miles) is taken into account, then human population density increases to 45.3 per km^2 (117.2 per sq. mile). This calculation includes all continental and island land area, including Antarctica. If Antarctica is also excluded, then population density rises to 50 people per km^2 (129.28 per sq. mile).[1] Considering that over half of the Earth's land mass consists of areas inhospitable to human inhabitation, such as deserts and high mountains, and that population tends to cluster around seaports and fresh water sources, this number by itself does not give any meaningful measurement of human population density.

Several of the most densely populated territories in the world are city-states, microstates, or dependencies.[] [5] These territories share a relatively small area and a high urbanization level, with an economically specialized city population drawing also on rural resources outside the area, illustrating the difference between high population density and overpopulation.

Mongolia is the least densely populated country in the world.

Cities with high population densities are, by some, considered to be overpopulated, though the extent to which this is the case depends on factors like quality of housing and infrastructure and access to resources.[6] Most of the most densely populated cities are in southern and eastern Asia, though Cairo and Lagos in Africa also fall into this category.[7]

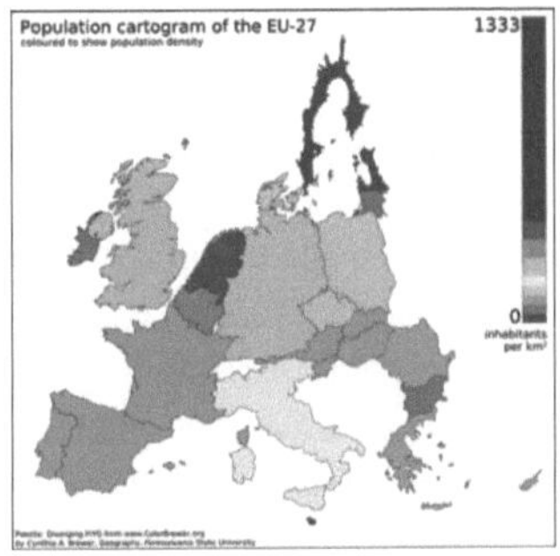

A population cartogram uses areas rather than colors to represent population.

City population is, however, heavily dependent on the definition of "urban area" used: densities are often higher for the central municipality itself, than when more recently developed and administratively separate suburban communities are included, as in the concepts of agglomeration or metropolitan area, the latter including sometimes neighboring cities. For instance, Milwaukee has a greater population density when just the inner city is measured, and not the surrounding suburbs as well.[8]

As a comparison, based on a world population of seven billion, the world's inhabitants would, as a loose crowd taking up ten square feet (one square metre) per person (Jacobs Method), would occupy a space a little larger than Delaware's land area.

Most densely populated countries

With population above 1 million

Rank	Country/Region	Population	Area (km^2)	Density (Pop. per km^2)
1	Singapore	5,183,700	710	7301
2	Hong Kong	7,061,200	1,104	6396
3	Bahrain	1,234,596	750	1646
4	Bangladesh	152,518,015	147,570	1034
5	Taiwan (R.O.C)	22,955,395	36,190	634
6	Mauritius	1,288,000	2,040	631
7	South Korea	48,456,369	99,538	487
8	Rwanda	10,718,379	26,338	407
9	Netherlands	16,760,000	41,526	404
10	Lebanon	4,224,000	10,452	404

With population above 10 million

Rank	Country/Region	Population	Area (km^2)	Density (Pop. per km^2)
1	Bangladesh	152,518,015	147,570	1034
2	Taiwan (R.O.C)	22,955,395	36,190	634
3	South Korea	48,456,369	99,538	487
4	Rwanda	10,718,379	26,338	407
5	Netherlands	16,760,000	41,526	404
6	India	1,210,193,422	3,185,263	380
7	Belgium	11,007,020	30,528	361
8	Japan	127,960,000	377,944	339
9	Sri Lanka	20,653,000	65,610	345

Other methods of measurement

While arithmetic density is the most common way of measuring population density, several other methods have been developed which aim to provide a more accurate measure of population density over a specific area.

- **Arithmetic density**: The total number of people / area of land (measured in km^2 or sq miles).
- **Physiological density**: The total population / area of arable land.
- **Agricultural density**: The total *rural* population / area of arable land.
- **Residential density** : The number of people living in an urban area / area of residential land.
- **Urban density** : The number of people inhabiting an urban area / total area of urban land.
- **Ecological optimum**: The density of population which can be supported by the natural resources.

See also

- Demography
- Human geography
- Idealized population
- Optimum population
- Population bottleneck
- Population genetics
- Population health
- Population momentum
- Population pyramid
- Rural transport problem
- Small population size
- Distance sampling

Lists:

- List of cities by population
- List of cities by population density
- List of European cities proper by population density
- List of islands by population density
- List of countries by population density
- List of U.S. states by population density

References

[1] Matt Rosenberg Population Density (http://geography.about.com/od/populationgeography/a/popdensity.htm). Geography.about.com. March 2 2011. Retrieved on 2011-12-10.

[2] Minimum viable population size (http://www.eoearth.org/article/Minimum_viable_population_size). Eoearth.org (2010-03-06). Retrieved on 2011-12-10.

[3] U.S. & World Population Clocks (http://www.census.gov/main/www/popclock.html). Census.gov. Retrieved on 2011-12-10.

[4] World (https://www.cia.gov/library/publications/the-world-factbook/geos/xx.html). CIA World Handbook

[5] The Monaco government uses a smaller surface area figure resulting in a population density of 18,078 per km^2

[6] Human Population (http://www.globalissues.org/issue/198/human-population). Global Issues. Retrieved on 2011-12-10.

[7] The largest cities in the world by land area, population and density (http://www.citymayors.com/statistics/largest-cities-density-125.html). Citymayors.com. Retrieved on 2011-12-10.

[8] The Population of Milwaukee County (http://www.wisconline.com/greenmap/milwaukee/population.html). Wisconline.com. Retrieved on 2011-12-10.

External links

- Selected Current and Historic City, Ward & Neighborhood Densities (http://www.demographia.com/db-citydenshist.htm)

Census

A **census** is the procedure of systematically acquiring and recording information about the members of a given population. It is a regularly occurring and official count of a particular population.[1] The term is used mostly in connection with national population and housing censuses; other common censuses include agriculture, business, and traffic censuses. The United Nations defines the essential features of population and housing censuses as "individual enumeration, universality within a defined territory, simultaneity and defined periodicity", and recommends that population censuses be taken at least every 10 years. United Nations recommendations also cover census topics to be collected, official definitions, classifications and other useful information to coordinate international practice.[2] [3]

Census taker visits a family living in a caravan, Netherlands 1925

The word is of Latin origin; during the Roman Republic, the census was a list that kept track of all adult males fit for military service. The modern census is essential to international comparisons of any kind of statistics and censuses collect data on many attributes of the population, not just how many people there are, although population estimates remain an important function of the census.

The census can be contrasted with sampling in which information is obtained only from a subset of a population, sometimes as an intercensal estimate. Census data are commonly used for research, business marketing, and planning, as well as a baseline for sampling surveys. Census counts are necessary to adjust samples to be representative of a population by weighting them as is common in opinion polling. Similarly, stratification requires knowledge of the relative sizes of different population strata which are derived from census enumerations. In some countries, census data are used to apportion electoral representation (sometimes controversially – e.g., *Utah v. Evans*).

Sampling

A census is often construed as the opposite of a sample as everyone is surveyed rather than a fraction. However, population census relies on a sampling frame to count the population. This is the only way to be sure that everyone has been included as otherwise those not responding would not be followed up and individuals could be missed. The fundamental premise of a census is that the population is not known and a new estimate is to be made by the analysis of primary data.

The use of a sampling frame is counterintuitive as it suggests that the population size is already known. However, a census is also used to collect attribute data on the individuals in the nation. This process of sampling marks the difference between historical census, which was a house to house process or the product of an imperial decree, and the modern statistical project.

The sampling frame used by census is almost always an address register. Thus it is not known if there is anyone resident or how many people there are in each household. Depending on the mode of enumeration, a form is sent to the householder, an enumerator calls, or administrative records for the dwelling are accessed. As a preliminary to the dispatch of forms, census workers will check any address problems on the ground.

While it may seem straightforward to use the postal service file for this purpose, this can be out of date and some dwellings may contain a number of independent households. A particular problem is what are termed 'communal establishments' which may be student residences, religious orders, homes for the elderly etc. As these are not easily enumerated by a single householder, they are often treated differently and visited by special teams of census workers to make sure they are classified appropriately.

Residence definitions

Individuals are normally counted within households and information is typically collected about the household structure and the housing. For this reason international documents refer to censuses of population and housing. Normally the census response is made by a household, indicating details of individuals resident there.

An important aspect of census enumerations is determining which individuals can be counted from which cannot be counted. Broadly, three definitions can be used: *de facto* residence; *de jure* residence; and, permanent residence. This is important to consider individuals who have multiple or temporary addresses. Every person should be identified uniquely as resident in one place but where they happen to be on Census Day, their *de facto* residence, may not be the best place to count them. Where an individual uses services may be more useful and this is at their usual, or *de jure*, residence. An individual may be represented at a permanent address, perhaps a family home for students or long term migrants.

It is necessary to have a precise definition of residence to decide whether visitors to a country should be included in the population count. This is becoming more important as students travel abroad for education for a period of several years. Other groups causing problems of enumeration are new born babies, refugees, people away on holiday, people moving home around census day, and people without a fixed address.

People having second homes because of working in another part of the country or retaining a holiday cottage are difficult to fix at a particular address sometimes causing double counting or houses being mistakenly identified as vacant. Another problem is where people use a different address at different times e.g. students living at their place of education in term time but returning to a family home during vacations or children whose parents have separated who effectively have two family homes. Census enumeration has always been based on finding people where they live as there is no systematic alternative - any list you could use to find people is derived from census activities in the first place. Recent UN guidelines provide recommendation on enumerating such complex households.[4]

Enumeration strategies

Historical censuses used crude enumeration assuming absolute accuracy. Modern approaches take into account the problems of overcount and undercount, and the coherence of census enumerations with other official sources of data.[5] This reflects a realist approach to measurement, acknowledging that under any definition of residence there is a true value of the population but this can never be measured with complete accuracy. An important aspect of the census process is to evaluate the quality of the data.[6]

Many countries use a post-enumeration survey to adjust the raw census counts.[7] This works in a similar manner to capture-recapture estimation for animal populations. In census circles this method is called dual system enumeration (DSE). A sample of households are visited by interviewers who record the details of the household as at census day. These data are then matched to census records and the number of people missed can be estimated by considering the number missed in the census or survey but counted in the other. This way counts can be adjusted for non-response varying between different demographic groups. An explanation using a fishing analogy can be found in "Trout, Catfish and Roach..."[8] which won an award from the Royal Statistical Society for excellence in official statistics in 2011.

Triple system enumeration has been proposed as an improvement as it would allow evaluation of the statistical dependence of pairs of sources. However, as the matching process is the most difficult aspect of census estimation this has never been implemented for a national enumeration. It would also be difficult to identify three different sources that were sufficiently different to make the triple system effort worthwhile. The DSE approach has another weakness in that it assumes there is no person counted twice (over count). In *de facto* residence definitions this would not be a problem but in *de jure* definitions individuals risk being recorded on more than one form leading to double counting. A particular problem here are students who often have a term time and family address.

Several countries have used a system which is known as short form/long form.[9] This is a sampling strategy which randomly chooses a proportion of people to send a more detailed questionnaire to (the long form). Everyone receives the short form questions. Thereby more data are collected but not imposing a burden on the whole population. This also reduces the burden on the statistical office. Indeed in the UK all residents were required to fill in the whole form but only a 10% sample were coded and analysed in detail, until 2001.[10] New technology means that all data are now scanned and processed. Recently there has been controversy in Canada about the cessation of the long form with the head, Munir Sheikh resigning.[11]

The use of alternative enumeration strategies is increasing[12] but these are not so simple as many people assume. The Netherlands has been most advanced in adopting a census using administrative data. This allows a simulated census to be conducted by linking several different administrative databases at an agreed time. Data can be matched and an overall enumeration established accounting for where the different sources are discrepant. A validation survey is still conducted in a similar way to the post enumeration survey employed in a traditional census.

Other countries which have a population register use this as a basis for all the census statistics needed by users. This is most common amongst Nordic countries but requires a large number of different registers to be combined including population, housing, employment and education. These registers are then combined and brought up to the standard of a statistical register by comparing the data in different sources and ensuring the quality is sufficient for official statistics to be produced.[13]

A recent innovation is the French instigation of a rolling census programme with different regions enumerated each year such that the whole country is completely enumerated every 5 years.

In Europe, in connection with the 2010 census round, a large number of countries adopted alternative census methodologies, often based on the combination of data from registers, surveys and other sources.[14]

Technology

Censuses have evolved in their use of technology with the latest censuses, the 2010 round, using many new types of computing. In Brazil, handheld devices were used by enumerators to locate residences on the ground. In many countries, census returns could be made via the Internet as well as in paper form. DSE is facilitated by computer matching techniques which can be automated, such as propensity score matching. In the UK, all census formats are scanned and stored electronically before being destroyed, replacing the need for physical archives. The record linking to perform an administrative census would not be possible without large databases being stored on computer systems.

New technology is not without problems in its introduction. The US census had intended to use the handheld computers but cost escalated and this was abandoned, with the contract being sold to Brazil. Online response is a good idea but one of the functions of census is to make sure everyone is counted accurately. A system which allowed people to enter their address without verification would be open to abuse. Therefore households have to be verified on the ground, typically by an enumerator visit or post out. Paper forms are still necessary for those without access to Internet connections. It is also plausible that the hidden nature of an administrative census means that users are not engaged with the importance of contributing their data to official statistics.

Uses of Census data

In the nineteenth century, the first censuses collected paper enumerations that had to be collated by hand so the statistical uses were very basic. The government owned the data and were able to publish statistics themselves on the state of the nation. Uses were to measure changes in the population and apportion representation. Population estimates could be compared to those of other countries.

By the beginning of the twentieth century, censuses were recording households and some indications of their employment. In some countries, census archives are released for public examination after many decades, allowing genealogists to track the ancestry of interested people. Archives provide a substantial historical record which may challenge established notions of tradition. It is also possible to understand the societal history through job titles and arrangements for the destitute and sick.

As governments assumed responsibility for schooling and welfare, large government departments made extensive use of census data. Actuarial estimates could be made to project populations and plan for provision in local government and regions. It was also possible for central government to allocate funding on the basis of census data.

Even into the mid twentieth century, census data was only directly accessible to large government departments. However, computers meant that tabulations could be used directly by university researchers, large businesses and local government offices. They could use the detail of the data to answer new questions and add to local and specialist knowledge.

Now, census data are published in a wide variety of formats to be accessible to business, all levels of governance, media, students and teachers, charities and researchers, and any citizen who is interested. Data can be represented visually or analysed in complex statistical models, to show the difference between certain areas, or to understand the association between different personal characteristics. Census data offer a unique insight into small areas and small demographic groups which sample data would be unable to capture with precision.

Privacy

Although the census provides a useful way of obtaining statistical information about a population, such information can sometimes lead to abuses, political or otherwise, made possible by the linking of individuals' identities to anonymous census data.[15] This consideration is particularly important when individuals' census responses are made available in microdata form, but even aggregate-level data can result in privacy breaches when dealing with small areas and/or rare subpopulations.

For instance, when reporting data from a large city, it might be appropriate to give the average income for black males aged between 50 and 60. However, doing this for a town that only has two black males in this age group would be a breach of privacy because either of those persons, knowing his own income and the reported average, could determine the other man's income.

Typically, census data are processed to obscure such individual information. Some agencies do this by intentionally introducing small statistical errors to prevent the identification of individuals in marginal populations;[16] others swap variables for similar respondents. Whatever measures have been taken to reduce the privacy risk in census data, new technology in the form of better electronic analysis of data poses increasing challenges to the protection of sensitive individual information. This known as statistical disclosure control.

Another possibility is to present survey results by means of statistical models in the form of a multivariate distribution mixture.[17] The statistical information in the form of conditional distributions (histograms) can be derived interactively from the estimated mixture model without any further access to the original database. As the final product does not contain any protected microdata, the model based interactive software [18] can be distributed without any confidentiality concerns.

Another method is simply to release no data at all, except very large scale data directly to the central government. Different release strategies between government have led to an international project (IPUMS) to co-ordinate access to microdata and corresponding metadata. Such projects also promote standardising metadata by projects such as SDMX so that best use can be made of the minimal data available.

Historical examples

Egypt

Censuses in Egypt are said to have been taken during the early Pharaonic period in 3340 BCE and in 3056 BCE.

Ancient Greece

There are several accounts of ancient Greek and Mesopotamian city states carrying out censuses.[19] The question of which is first is clouded by very different approaches: counting only men, counting a pile of rocks etc. but such censuses took place 1600 BCE and earlier.

Ancient Israel

Censuses are mentioned in the Bible. God commands a flat tax to be paid with the census in Exodus 30:11-16 for the upkeep of the Tabernacle. The Book of Numbers is named after the counting of the Israelite population (in Numbers 1-4) according to the house of the Fathers after the exodus from Egypt. A second census was taken while the Israelite were camped in the plains of Moab, in Numbers 26.

King David performed a census that produced disastrous results (in Samuel 2 24 and Chronicles 1 21). His son, King Solomon, had all of the foreigners in Israel counted in Chronicles 2 2:17.

The Gospel of Luke records Jesus being born during a census in Luke 2.

China

In 2 CE during the Han Dynasty, China held a census still considered by scholars to be quite accurate.[20] [21] The census is considered one of the world's earliest preserved censuses,[] and found 57.67 million people registered in 12.36 million households.[22] [23] The areas with the highest population densities were the Yellow River and Huai River valleys. Chengdu was the largest city in China, reaching a population of 282,147 people.[] Another census dates to AD 144, when 49.73 million people were recorded living in 9.94 million households.

India

The oldest recorded census in India is thought to have occurred around 300 BCE during the reign of the Emperor Chandragupta Maurya under the leadership of Kautilya or Chanakya.[24]

Rome

The word "census" originated in ancient Rome from the Latin word *censere* ("to estimate"). The census played a crucial role in the administration of the Roman Empire, as it was used to determine taxes. With few interruptions, it was usually carried out every five years.[25] It provided a register of citizens and their property from which their duties and privileges could be listed. It is said to have been instituted by the Roman king Servius Tullius in the 6th century BCE,[26] at which time the number of arms-bearing citizens was counted at around 80,000.[27]

Rashidun and Umayyad Caliphates

In the Middle Ages, the Caliphate began conducting regular censuses soon after its formation, beginning with the one ordered by the second Rashidun caliph, Umar.[28]

Medieval Europe

The Domesday Book was undertaken in 1086 CE by William I of England so that he could properly tax the land he had recently conquered in medieval Europe. In 1183 CE, a census was taken of the crusader Kingdom of Jerusalem, to ascertain the number of men and amount of money that could possibly be raised against an invasion by Saladin, sultan of Egypt and Syria.

Inca Empire

In the 15th century, the Inca Empire had a unique way to record census information. The Incas did not have any written language but recorded information collected during censuses and other numeric information as well as non-numeric data on quipus, strings from llama or alpaca hair or cotton cords with numeric and other values encoded by knots in a base-10 positional system.

Spanish empire

On May 25, 1577, King Philip II of Spain ordered by royal cédula the preparation of a general description of Spain's holdings in the Indies. Instructions and a questionnaire, issued in 1577 by the Office of the Cronista Mayor, were distributed to local officials in the Viceroyalties of New Spain and Peru to direct the gathering of information. The questionnaire, composed of fifty items, was designed to elicit basic information about the nature of the land and the life of its peoples. The replies, known as "relaciones geográficas," were written between 1579 and 1585 and were returned to the Cronista Mayor in Spain by the Council of the Indies.

See also

- IBM and the Holocaust
- *Liber Censuum*
- Race and ethnicity in censuses
- Social research

Notes

[2] United Nations (2008). Principles and Recommendations for Population and Housing Censuses (http://unstats.un.org/unsd/demographic/sources/census/docs/P&R_Rev2.pdf). Statistical Papers: Series M No. 67/Rev.2. p8. ISBN 978-92-1-161505-0.

[3] http://www.unece.org/fileadmin/DAM/stats/publications/CES_2010_Census_Recommendations_English.pdf

[6] Leo Breiman (1994). "The 1991 Census Adjustment: Undercount or Bad Data?", *Statistical Science* **9** (4), pp.458-475.

[7] World Population and Housing Census Programme (2010) Post Enumeration Surveys: Operational guidelines (http://unstats.un.org/unsd/demographic/standmeth/handbooks/Manual_PESen.pdf), United Nations Secretariat, Dept of Economic and Social Affairs, Statistics Division, Tech Report

[8] Benton, P. Trout, Catfish and Roach: The beginner's guide to census population estimates (http://www.ons.gov.uk/ons/guide-method/census/2011/the-2011-census/census-coverage-survey/trout--catfish-and-roach.pdf), Office for National Statistics, UK

[9] Other methods of census taking (http://www.ons.gov.uk/ons/guide-method/census/2011/what-is-a-census/other-methods-of-census-taking/index.html), Office for National Statistics, UK

[15] The Census and Privacy (http://www.epic.org/privacy/census/).

[18] http://ro.utia.cas.cz/census

[21] Twitchett, D., Loewe, M., and Fairbank, J.K. *Cambridge History of China: The Ch'in and Han Empires 221 B.C.-A.D. 220*. Cambridge University Press (1986), p. 240.

[22] Nishijima (1986), 595–596.

[23] H. Yoon (1985). "An early Chinese idea of a dynamic environmental cycle", *GeoJournal* **10** (2), 211-212.

[25] Scheidel, Walter (2009) *Rome and China: comparative perspectives on ancient world empires*. Oxford University Press, p. 28.

[26] Livy *Ab urbe condita* 1.42

[27] Livy *Ab urbe condita* 1.42, citing Fabius Pictor

References

- Alterman, Hyman, (1969). *Counting People: The Census in History*. Harcourt, Brace & Company.
- Bielenstein, Hans, (1978). "Wang Mang, the restoration of the Han dynasty, and Later Han." In *The Cambridge History of China*, vol. 1, eds. Denis Twitchett and John K. Fairbank, p. 223-90, Cambridge: Cambridge University Press.
- Krüger, Stephen, (Fall 1991). "The Decennial Census", 19 *Western State University Law Review* 1 (http://wsulawreview.org/Volume19Articles.pdf); available at HeinOnline (http://heinonline.org/HOL/LandingPage?collection=journals&handle=hein.journals/wsulr19&div=6&id=&page=) (subscription required).
- Effects of UK 'Jedi' hoax on 2001 UK census from ONS (http://www.statistics.gov.uk/cci/nugget.asp?id=297).
- U.S. Census Press Release on 1930 Census (http://www.census.gov/Press-Release/www/releases/archives/miscellaneous/000507.html).
- U.S. Census Press Release on Soundex and WPA (http://www.archives.gov/publications/prologue/spring_2002_soundex_projects.html).

External links

- My Free Census (http://www.myfreecensus.com/) Find your ancestors in the U.S. census, and International census, with these free online census records, census images, census indexes and census databases in our easy to use census directory.
- Census of Ireland 1911 (http://www.census.nationalarchives.ie/).
- Online Historical Population Reports Project (OHPR) (http://www.histpop.org).
- PR as a function of census management: comparative analysis of fifteen census experiences (http://www.tesionline.com/intl/thesis.jsp?idt=8384)

List_of_highest-income_places_in_the_United_States

The **highest-income places in the United States** are mostly found in metropolitan areas such as New York City, San Francisco, Los Angeles, Chicago, Washington, D.C., Boston, and Miami with a high density of wealthy individuals.

Highest-income places by per capita income[1]

100 highest-income places with at least 1,000 households

This is a list of the 100 richest places in the United States with at least 1,000 households. This is the most common list used for referring to the richest communities in the country, as it eliminates any places with insignificant populations. It is usually preferred over a list of places with a population of 1,000, because it means there are at least 1,000 bases for the income statistics.

Number of places by state in the top 100: California 23, New York 11, Illinois 9, Florida 9, Connecticut 6, New Jersey 6, Texas 6, Colorado 4, Michigan 4, Ohio 3, Maryland 2, Minnesota 2, Missouri 2, Pennsylvania 2, Tennessee 2, Washington 2, Arizona 1, Delaware 1, Kansas 1, Kentucky 1, Massachusetts 2, Oklahoma 1, Virginia 1. Twenty-seven states do not have any places in the top 100.

Rank	Place	Per Capita Income	Number of Households
1	Rancho Santa Fe, California	$113,132	1,204
2	Bloomfield Hills, Michigan	$104,920	1,489
3	Belle Meade, Tennessee	$104,908	1,067
4	Woodside, California	$104,667	1,949
5	Indian River Shores, Florida	$102,511	1,854
6	Cherry Hills Village, Colorado	$99,996	1,980
7	Portola Valley, California	$99,621	1,700
8	Newport Coast, California	$98,770	1,001
9	Hillsborough, California	$98,643	3,689
10	Piney Point Village, Texas	$97,247	1,225
11	Highland Park, Texas	$97,008	3,585
12	The Village of Indian Hill, Ohio	$96,872	2,066
13	Mission Hills, Kansas	$95,405	1,284

14	Los Altos Hills, California	$92,840	2,740
15	Gladwyne, Pennsylvania	$90,940	1,476
16	South Beach, Florida	$90,938	1,610
17	Armonk, New York	$89,907	1,172
18	Scarsdale, New York	$89,907	2,413
19	Ladue, Missouri	$89,623	3,414
20	Bronxville, New York	$89,483	2,312
21	Pelican Bay, Florida	$89,063	3,005
22	Hunters Creek Village, Texas	$88,821	1,471
23	Glencoe, Illinois	$88,059	3,072
24	Muttontown, New York	$88,020	1,022
25	Bunker Hill Village, Texas	$86,434	1,226
26	Tiburon, California	$85,966	3,712
27	Saddle River, New Jersey	$85,934	1,118
28	Winnetka, Illinois	$84,134	4,162
29	Greenville, Delaware	$83,223	1,190
30	New Canaan, Connecticut	$100,824	7,010
31	Medina, Washington	$81,742	1,111
32	Paradise Valley, Arizona	$81,290	5,034
33	North Barrington, Illinois	$81,243	1,003
34	Sausalito, California	$81,040	4,254
35	Longboat Key, Florida	$80,963	4,280
36	Fox Chapel, Pennsylvania	$80,610	1,875
37	Weston, Massachusetts	$79,640	3,718
38	Genesee, Colorado	$79,180	1,511
39	Clyde Hill, Washington	$78,252	1,054
40	Great Falls, Virginia	$78,149	2,785
41	Chappaqua, New York	$77,835	3,118
42	Darien, Connecticut	$95,577	6,698
43	Travilah, Maryland	$77,129	2,366
44	Lake Forest, Illinois	$77,092	6,687
45	Alpine, New Jersey	$76,995	1,204
46	Millburn, New Jersey	$76,796	7,015
47	Lloyd Harbor, New York	$76,696	1,147
48	Oak Brook, Illinois	$76,688	3,124
49	Monte Sereno, California	$76,577	1,211
50	Rye (city), New York	$76,566	5,377
51	Indian Wells, California	$76,187	1,982
52	South Barrington, Illinois	$76,078	1,147

53	Carlisle, Massachusetts	$75,553	1,618
54	Pound Ridge, New York	$74,914	1,699
55	Weston, Connecticut	$92,735	3,379
56	Greenwich, Connecticut	$92,759	23,076
57	Malibu, California	$74,336	5,137
58	New Castle, New York	$73,888	5,732
59	Rumson, New Jersey	$73,692	2,452
60	Larchmont, New York	$73,675	2,418
61	Westport, Connecticut	$90,792	9,573
62	Nichols Hills, Oklahoma	$73,661	1,729
63	Barrington Hills, Illinois	$73,629	1,381
64	Inverness, Illinois	$73,271	2,312
65	Harding Township, New Jersey	$72,689	1,180
66	North Oaks, Minnesota	$72,686	1,300
67	Moreland Hills, Ohio	$72,001	1,286
68	Kings Point, New York	$71,867	1,401
69	Pepper Pike, Ohio	$71,225	2,203
70	Franklin, Michigan	$71,033	1,073
71	Del Monte Forest, California	$70,609	2,092
72	Piedmont, California	$70,539	3,804
73	Castle Pines, Colorado	$70,456	2,078
74	Old Westbury, New York	$70,089	1,063
75	Montecito, California	$70,077	3,686
76	Bernardsville, New Jersey	$69,854	2,723
77	Grosse Pointe Township, Michigan	$69,731	1,018
78	West University Place, Texas	$69,674	5,286
79	Grosse Pointe Shores, Michigan	$69,639	1,058
80	Town and Country, Missouri	$69,347	3,593
81	Greenwood Village, Colorado	$69,189	3,997
82	Palos Verdes Estates, California	$69,040	4,993
83	Emerald Lake Hills, California	$68,966	1,437
84	Loyola, California	$68,730	1,275
85	Forest Hills, Tennessee	$68,228	1,729
86	Riverwoods, Illinois	$67,878	1,261
87	Bal Harbour, Florida	$67,680	1,908
88	Highland Beach, Florida	$67,288	2,192
89	Blackhawk-Camino Tassajara, California	$66,972	3,326
90	Sanibel, Florida	$66,912	3,049
91	Boca Pointe, Florida	$66,797	1,824

92	Los Altos, California	$66,776	10,462
93	Indian Hills, Kentucky	$66,637	1,119
94	San Joaquin Hills, California	$66,544	1,376
95	Brookmont, Maryland	$66,465	1,194
96	Orono, Minnesota	$65,825	2,766
97	Wilton, Connecticut	$65,806	5,923
98	Alamo, California	$65,705	5,406
99	Olmos Park, Texas	$65,697	1,043
100	Johns Creek, Georgia	$65,994	23,013

100 highest-income places with a population of at least 1,000

Number of places by state in the top 100: New York 21, California 16, Illinois 9, Florida 7, New Jersey 6, Texas 6, Connecticut 5, Michigan 4, Ohio 4, Colorado 2, Washington 3, Maryland 3, Pennsylvania 2, Massachusetts 2, Arizona 1, Delaware 1, Kansas 1, Maine 1, Minnesota 1, Missouri 1, North Carolina 1, Oklahoma 1, Tennessee 1, Virginia 1, Wisconsin 1. Twenty-five states do not have any places in the top 100.

Rank	Place	Per Capita Income	Population
1	Belvedere, California	$113,595	2,125
2	Rancho Santa Fe, California	$113,132	3,252
3	Atherton, California	$112,408	7,194
4	Rolling Hills, California	$111,031	1,871
5	Barton Creek, Texas	$110,504	1,589
6	Palm Beach, Florida	$109,219	10,468
7	Bloomfield Hills, Michigan	$104,920	3,940
8	Belle Meade, Tennessee	$104,908	2,943
9	Woodside, California	$104,667	5,352
10	Oyster Bay Cove, New York	$103,203	2,262
11	Indian River Shores, Florida	$102,511	3,448
12	Kenilworth, Illinois	$100,718	2,494
13	North Hills, New York	$100,093	4,301
14	Cherry Hills Village, Colorado	$99,996	5,958
15	Portola Valley, California	$99,621	4,462
16	Newport Coast, California	$98,770	2,671
17	Hillsborough, California	$98,643	10,825
18	Piney Point Village, Texas	$97,247	3,380
19	Highland Park, Texas	$97,008	8,842
20	The Village of Indian Hill, Ohio	$96,872	5,907
21	Sands Point, New York	$95,647	2,786
22	Mission Hills, Kansas	$95,405	3,593

23	Chevy Chase Village, Maryland	$95,174	2,043
24	Plandome, New York	$95,102	1,272
25	River Hills, Wisconsin	$94,479	1,631
26	Fairbanks Ranch, California	$94,150	2,244
27	Hidden Hills, California	$94,096	1,875
28	Los Altos Hills, California	$92,840	7,902
29	Gladwyne, Pennsylvania	$90,940	4,050
30	South Beach, Florida	$90,938	3,457
31	Scarsdale, New York	$89,907	17,823
32	Ladue, Missouri	$89,623	8,645
33	Bronxville, New York	$89,483	6,543
34	Pelican Bay, Florida	$89,063	5,686
35	Hunters Creek Village, Texas	$88,821	4,374
36	Glencoe, Illinois	$88,059	8,762
37	Muttontown, New York	$88,020	3,412
38	Bunker Hill Village, Texas	$86,434	3,654
39	Tiburon, California	$85,966	8,666
40	Saddle River, New Jersey	$85,934	3,201
41	Biltmore Forest, North Carolina	$85,044	1,440
42	Brookville, New York	$84,375	2,126
43	Winnetka, Illinois	$84,134	12,419
44	Laurel Hollow, New York	$83,366	1,930
45	Greenville, Delaware	$83,223	2,332
46	North Key Largo, Florida	$83,199	1,049
47	Somerset, Maryland	$82,368	1,124
48	Hewlett Harbor, New York	$82,069	1,271
49	New Canaan, Connecticut	$82,049	19,395
50	Medina, Washington	$81,742	3,011
51	Paradise Valley, Arizona	$81,290	13,664
52	North Barrington, Illinois	$81,243	2,918
53	Sausalito, California	$81,040	7,330
54	Longboat Key, Florida	$80,963	7,603
55	Fox Chapel, Pennsylvania	$80,610	5,436
56	Weston, Massachusetts	$79,640	11,469
57	Genesee, Colorado	$79,180	3,699
58	Clyde Hill, Washington	$78,252	2,890
59	Great Falls, Virginia	$78,149	8,549
60	Old Brookville, New York	$77,874	2,167
61	Chappaqua, New York	$77,835	9,468

62	Darien, Connecticut	$77,519	19,607
63	Essex Fells, New Jersey	$77,434	2,162
64	Hill Country Village, Texas	$77,374	1,028
65	Travilah, Maryland	$77,129	7,442
66	Lake Forest, Illinois	$77,092	20,059
67	Alpine, New Jersey	$76,995	2,183
68	Millburn, New Jersey	$76,796	19,795
69	Lloyd Harbor, New York	$76,696	3,675
70	Oak Brook, Illinois	$76,668	8,702
71	Monte Sereno, California	$76,577	3,483
72	Rye (city), New York	$76,566	14,955
73	Lattingtown, New York	$76,260	1,860
74	Indian Wells, California	$76,187	3,816
75	Ocean Ridge, Florida	$76,088	1,636
76	South Barrington, Illinois	$76,078	3,760
77	Carlisle, Massachusetts	$75,553	4,852
78	Pound Ridge, New York	$74,914	4,726
79	Weston, Connecticut	$74,817	10,037
80	Bingham Farms, Michigan	$74,588	1,030
81	Greenwich, Connecticut	$74,346	61,101
82	Malibu, California	$74,336	13,000
83	New Castle, New York	$73,888	17,491
84	Rumson, New Jersey	$73,692	7,137
85	Wetherington, Ohio	$73,684	1,010
86	Larchmont, New York	$73,675	6,485
87	Westport, Connecticut	$73,664	25,749
88	Nichols Hills, Oklahoma	$73,661	4,056
89	Barrington Hills, Illinois	$73,629	3,915
90	Roslyn Estates, New York	$73,628	1,210
91	Inverness, Illinois	$73,271	6,749
92	Bratenahl, Ohio	$72,757	1,337
93	Harding Township, New Jersey	$72,689	3,180
94	North Oaks, Minnesota	$72,686	3,883
95	Great Neck Estates, New York	$72,476	2,756
96	Cape Elizabeth, Maine	$72,359	9,068
97	Yarrow Point, Washington	$72,135	1,008
98	Moreland Hills, Ohio	$72,001	3,298
99	Kings Point, New York	$71,867	5,076
100	Wayland, Massachusetts	$71,798	13,900

100 highest-income places in the United States

This is a list of the 100 richest places in the United States, regardless of population or number of households (US Census 2000).

Ranking of states with places in the top 100, from most to least: New York 16, Florida 15, California 14, Texas 6, Illinois 5, Missouri 5, Pennsylvania 4, Michigan 3, Minnesota 3, New Jersey 3, Wisconsin 3, Delaware 2, Indiana 2, Kansas 2, Kentucky 2, Maryland 2, North Carolina 2, Ohio 2, Washington 2, Alaska 1, Arizona 1, Colorado 1, Connecticut 1, Louisiana 1, Massachusetts 1, Tennessee 1. Twenty-four states do not have any places in the top 100.

Rank	Place	Per Capita Income	Population	Number of Households
1	Fisher Island, Florida	$236,238	467	218
2	Jupiter Island, Florida	$200,087	620	285
3	Baker, Missouri	$182,000	5	2
4	Rex, North Carolina	$148,073	55	17
5	Golf, Florida	$144,956	230	119
6	Hunting Valley, Ohio	$144,281	735	284
7	Manalapan, Florida	$143,729	321	167
8	Indian Creek, Florida**	$137,382	33	14
9	Orchid, Florida	$135,870	140	69
10	Mockingbird Valley, Kentucky	$134,745	190	74
11	Gulf Stream, Florida	$133,651	716	340
12	Westover Hills, Texas	$133,558	658	258
13	Green Hills, Pennsylvania	$124,279	18	7
14	Willowbrook, Kansas	$118,946	36	13
15	Mantoloking, New Jersey	$114,017	423	207
16	Hunts Point, Washington	$113,816	443	165
17	Belvedere, California	$113,595	2,125	956
18	Hewlett Bay Park, New York	$113,320	484	157
19	Rancho Santa Fe, California	$113,132	3,252	1,204
20	Atherton, California	$112,408	7,194	2,413
21	Rolling Hills, California	$111,031	1,871	645
22	Barton Hills, Michigan	$110,683	335	136
23	Barton Creek, Texas	$110,504	1,589	707
24	Cove Neck, New York	$110,139	300	110
25	Palm Beach, Florida	$109,219	10,468	5,789
26	Bloomfield Hills, Michigan	$104,920	3,940	1,520
27	Belle Meade, Tennessee	$104,908	2,943	1,067
28	Woodside, California	$104,667	5,352	1,949
29	Huntleigh, Missouri	$104,420	323	122
30	Oyster Bay Cove, New York	$103,203	2,262	725

31	Indian River Shores, Florida	$102,511	3,448	1,854
32	Kenilworth, Illinois	$100,718	2,494	792
33	Country Life Acres, Missouri	$100,617	81	27
34	Crows Nest, Indiana	$100,565	96	39
35	North Hills, New York	$100,093	4,301	1,808
36	Cherry Hills Village, Colorado	$99,996	5,958	1,980
37	Portola Valley, California	$99,621	4,462	1,700
38	Newport Coast, California	$98,770	2,671	1,001
39	Hillsborough, California	$98,643	10,825	3,689
40	Piney Point Village, Texas	$97,247	3,380	1,225
41	Highland Park, Texas	$97,008	8,842	3,585
42	The Village of Indian Hill, Ohio	$96,872	5,907	2,066
43	Centre Island, New York	$96,674	444	174
44	Sands Point, New York	$95,647	2,786	878
45	Woodland, Minnesota	$95,495	480	173
46	Diablo, California	$95,419	988	341
47	Mission Hills, Kansas	$95,405	3,593	1,284
48	Chevy Chase Village, Maryland	$95,174	2,043	704
49	Plandome, New York	$95,102	1,272	409
50	River Hills, Wisconsin	$94,479	1,631	590
51	Fairbanks Ranch, California	$94,150	2,244	734
52	Hidden Hills, California	$94,096	1,875	568
53	Matinecock, New York	$93,559	836	285
54	Los Altos Hills, California	$92,840	7,902	2,740
55	Mound, Louisiana	$92,200	12	4
56	Minnetonka Beach, Minnesota	$91,844	614	215
57	Gladwyne, Pennsylvania	$90,940	4,050	1,476
58	South Beach, Florida	$90,938	3,457	1,610
59	Scarsdale, New York	$89,907	17,823	5,662
60	Halibut Cove, Alaska	$89,895	35	18
61	Ladue, Missouri	$89,623	8,645	3,414
62	Bronxville, New York	$89,483	6,543	2,312
63	Mettawa, Illinois	$89,104	367	135
64	Pelican Bay, Florida	$89,063	5,686	3,005
65	Hunters Creek Village, Texas	$88,821	4,374	1,471
66	Glencoe, Illinois	$88,059	8,762	3,072
67	Hewlett Neck, New York	$88,049	504	155
68	Muttontown, New York	$88,020	3,412	1,022
69	Chenequa, Wisconsin	$86,552	583	223

70	Bunker Hill Village, Texas	$86,434	3,654	1,226
71	Oak Point, Florida	$86,225	145	48
72	Tiburon, California	$85,966	8,666	3,712
73	Saddle River, New Jersey	$85,934	3,201	1,118
74	Glenview, Kentucky	$85,094	558	214
75	Biltmore Forest, North Carolina	$85,044	1,440	588
76	Grand View-on-Hudson, New York	$84,707	284	132
77	Brookville, New York	$84,375	2,126	631
78	Winnetka, Illinois	$84,134	12,419	4,162
79	Lake Angelus, Michigan	$83,792	326	132
80	Laurel Hollow, New York	$83,366	1,930	598
81	Greenville, Delaware	$83,223	2,332	1,190
82	North Key Largo, Florida	$83,199	1,049	565
83	Somerset, Maryland	$82,368	1,124	406
84	Sunfish Lake, Minnesota	$82,347	504	173
85	Williams Creek, Indiana	$82,132	413	157
86	Henlopen Acres, Delaware	$82,091	139	69
87	Hewlett Harbor, New York	$82,069	1,271	429
88	New Canaan, Connecticut	$82,049	19,395	6,822
89	Medina, Washington	$81,742	3,011	1,111
90	Oconomowoc Lake, Wisconsin	$81,593	564	208
91	Far Hills, New Jersey	$81,535	859	368
92	Paradise Valley, Arizona	$81,290	13,664	5,034
93	North Barrington, Illinois	$81,243	2,918	1,003
94	Sausalito, California	$81,040	7,330	4,254
95	Westwood, Missouri	$80,990	284	120
96	Longboat Key, Florida	$80,963	7,603	4,280
97	Fox Chapel, Pennsylvania	$80,610	5,436	1,875
98	Weston, Massachusetts	$79,640	11,469	3,718
99	Sewickley Heights, Pennsylvania	$79,541	981	336
100	North Beach, Florida	$79,269	243	93
	**See City Article for Rank Explanation			

100 highest-income places with a population of at least 50,000

Rank		Population	Per Capita Income
1	Greenwich, Connecticut	61,101	74,346
2	Arcadia, California	56,364	67,400
3	Johns Creek, Georgia	76,728	65,994
4	Newport Beach, California	85,186	63,015
5	Bethesda, Maryland	70,032	58,479
6	Palo Alto, California	58,598	57,257
7	Newton, Massachusetts	83,829	45,708
8	Boca Raton, Florida	74,764	45,628
9	Sandy Springs, Georgia	85,782	45,494
10	West Bloomfield Township, Michigan	64,862	44,885
11	Cupertino, California	50,546	44,749
12	Brookline, Massachusetts	57,107	44,327
13	Naperville, Illinois	141,644	44,235
14	Fairfield, Connecticut	57,340	43,670
15	Santa Monica, California	84,084	42,874
16	Reston, Virginia	56,407	42,747
17	Catalina Foothills, Arizona	53,794	43,006
18	Pleasanton, California	63,654	41,623
19	Minnetonka, Minnesota	51,301	40,410
20	Walnut Creek, California	64,296	39,875
21	Mountain View, California	70,708	39,693
22	Laguna Niguel, California	61,891	39,167
23	Scottsdale, Arizona	202,705	39,158
24	Eden Prairie, Minnesota	54,901	38,854
25	Redondo Beach, California	63,261	38,305
26	The Woodlands, Texas	55,649	37,724
27	Arlington, Virginia	189,453	37,706
28	Alexandria, Virginia	128,283	37,645
29	Bellevue, Washington	109,569	36,905
30	Sunnyvale, California	131,760	36,524
31	Plano, Texas	222,030	36,514
32	Oak Park, Illinois	52,524	36,340
33	Plymouth, Minnesota	65,894	36,309
34	San Mateo, California	92,482	36,176
35	Yorba Linda, California	58,918	36,173
36	Farmington Hills, Michigan	82,111	36,134
37	Roswell, Georgia	79,334	36,012

38	Troy, Michigan	80,959	35,936
39	San Rafael, California	56,063	35,762
40	Naperville, Illinois	128,358	35,551
41	Weston, Florida	63,534	35,490
42	Wayne, New Jersey	54,069	35,349
43	Rochester Hills, Michigan	68,825	35,070
44	Stamford, Connecticut	117,083	34,987
45	Burke, Virginia	57,737	34,936
46	Carlsbad, California	78,247	34,863
47	Highlands Ranch, Colorado	70,931	34,707
48	Flower Mound, Texas	50,702	34,699
49	San Francisco, California	776,733	34,556
50	Encinitas, California	58,014	34,336
51	Thousand Oaks, California (includes Newbury Park, California)	117,005	34,314
52	Wheaton, Illinois	55,416	34,147
53	Redwood City, California	75,402	34,042
54	White Plains, New York	53,077	33,825
55	Evanston, Illinois	74,239	33,645
56	Arlington Heights, Illinois	76,031	33,544
57	Sugar Land, Texas	63,328	33,506
58	West Hartford, Connecticut	63,589	33,468
59	Hoover, Alabama	62,742	33,361
60	Ellicott City, Maryland	56,397	33,316
61	Mission Viejo, California	93,102	33,302
62	Cary, North Carolina	94,536	32,974
63	Columbia, Maryland	88,254	32,833
64	Towson, Maryland	51,793	32,502
65	Irvine, California	212,793	32,196
66	Overland Park, Kansas	149,080	32,069
67	Huntington Beach, California	189,594	31,964
68	New Rochelle, New York	72,182	31,956
69	Norwalk, Connecticut	82,951	31,781
70	Santa Clara, California	102,361	31,755
71	Annandale, Virginia	54,994	31,623
72	Fremont, California	203,413	31,411
73	Cambridge, Massachusetts	101,355	31,156
74	Livermore, California	73,345	31,062
75	Royal Oak, Michigan	60,062	30,990
76	Alameda, California	72,259	30,982

77	Bowie, Maryland	50,269	30,703
78	Palatine, Illinois	65,479	30,661
79	Schaumburg, Illinois	75,386	30,587
80	Maple Grove, Minnesota	50,365	30,544
81	Berkeley, California	102,743	30,477
82	Orland Park, Illinois	51,077	30,467
83	Castro Valley, California	57,292	30,454
84	Seattle, Washington	563,374	30,306
85	Folsom, California	51,884	30,210
86	Eagan, Minnesota	63,557	30,167
87	Edison, New Jersey	97,687	30,148
88	Shelby Charter Township, Michigan	65,159	30,131
89	Bloomington, Minnesota	85,172	29,782
90	Richardson, Texas	91,802	29,551
91	Delray Beach, Florida	60,020	29,350
92	Goleta, California	55,204	28,890
93	Milford, Connecticut	50,594	28,773
94	Washington, D.C.	572,059	28,659
95	Camarillo, California	57,077	28,635
96	Canton, Michigan	76,366	28,609
97	Lake Forest, California	58,707	28,583
98	Plantation, Florida	82,934	28,250
99	Pasadena, California	133,936	28,186
100	McKinney, Texas	54,369	28,185

Highest-income places by median household income[1]

Places with median household incomes in excess of $200,000

Note: Data for places with extremely small populations may be misleading.

Place	Population
Atherton, California	7,210
Brookville, New York	2,146
Chevy Chase Village, Maryland	2,051
Fairbanks Ranch, California	1,877
Fisher Island, Florida	339
Golf, Florida	239
Hewlett Bay Park, New York	485
Hidden Hills, California	1,875
Hunting Valley, Ohio	737

Huntleigh, Missouri	305
Jupiter Island, Florida	621
Kenilworth, Illinois	2,494
Laurel Hollow, New York	2,046
Mockingbird Valley, Kentucky	194
Mound, Louisiana	22
Orchid, Florida	138
Oyster Bay Cove, New York	2,184
Rancho Santa Fe, California	3,271
Rex, North Carolina	15
Rolling Hills, California	1,871
Sands Point, New York	2,761
Westover Hills, Texas	649

Highest-income places with a population of at least 10,000

Rank	Place	Population	Median Household Income
1	Scarsdale, New York	17,823	$182,792
2	New Canaan, Connecticut	19,395	$178,651
3	Winnetka, Illinois	12,388	$167,458
4	Darien, Connecticut	19,607	$160,274
5	Weston, Massachusetts	11,469	$153,918
6	Paradise Valley, Arizona	13,664	$150,228
7	Weston, Connecticut	10,037	$146,697
8	Wilton, Connecticut	17,633	$141,428
9	Town and Country, Missouri	10,879	$139,967
10	Saratoga, California	29,855	$139,895
11	Alamo, California	15,142	$137,105
12	Coto de Caza, California	13,057	$136,726
13	Lake Forest, Illinois	20,018	$136,462
14	Wolf Trap, Virginia	13,879	$135,782
15	West Windsor, New Jersey	27,165	$134,353
16	Franklin Lakes, New Jersey	10,422	$132,373
17	Southlake, Texas	21,497	$131,549
18	Millburn, New Jersey	19,765	$130,848
19	West University Place, Texas	14,211	$130,721
20	Burr Ridge, Illinois	10,328	$129,507
21	Potomac, Maryland	44,821	$128,936
22	Los Altos, California	27,585	$126,740
23	Wellesley, Massachusetts	26,613	$125,814

24	Westport, Connecticut	25,749	$119,872
25	Sudbury, Massachusetts	16,841	$118,579
26	Orinda, California	17,446	$117,637
27	Colleyville, Texas	19,574	$117,419
28	San Marino, California	12,973	$117,267
29	West Windsor, New Jersey	21,907	$116,335
30	Greenwood Village, Colorado	11,068	$116,147
31	Summit, New Jersey	21,103	$115,606
32	Concord, Massachusetts	16,993	115, 897
33	Danville, California	42,127	$114,064
34	Acton, Massachusetts	20,331	$112,532
35	Cinco Ranch, Texas	11,237	$111,517
36	Rye, New York	14,955	$110,894
37	Montecito, California	10,090	$110,669
38	La Canada Flintridge, California	20,381	$109,989
39	Colts Neck Township, New Jersey	12,331	$109,190
40	North Potomac, Maryland	22,945	$109,173
41	Berkeley Heights, New Jersey	13,407	$107,716
42	Pinecrest, Florida	19,181	$107,507
43	Ridgefield, Connecticut	23,643	$107,351
44	Deerfield, Illinois	18,497	$107,194
45	Wilmette, Illinois	27,684	$106,773
46	Cutler, Florida	17,656	$106,432
47	Hinsdale, Illinois	17,482	$104,551
48	Ridgewood, New Jersey	24,936	$104,286
49	Glen Rock, New Jersey	11,546	$104,192
50	Garden City, New York	21,662	$104,176
51	Dix Hills, New York	26,096	$104,160
52	Bloomfield Township, Michigan	43,027	$103,897
53	Wyckoff, New Jersey	16,508	$103,614
54	Milton, Massachusetts	27,003	$103,373
55	Parkland, Florida	13,988	$102,624
56	Leawood, Kansas	27,870	$102,496
57	Fort Hunt, Virginia	12,827	$102,259
57	Highland Village, Texas	12,163	$102,141
58	Lafayette, California	23,463	$102,107
59	Tamalpais-Homestead Valley, California	10,596	$102,094
60	Malibu, California	12,514	$102,031
61	Clayton, California	10,792	$101,651

62	Sammamish, Washington	34,119	$101,592
63	Jericho, New York	12,984	$101,477
64	Wayland, Massachusetts	13,100	$101,036
65	Highland Park, Illinois	31,379	$100,967
66	Hockessin, Delaware	12,896	$100,844
67	Manhattan Beach, California	34,039	$100,750
68	Mountain Brook, Alabama	20,983	$100,483
69	Cupertino, California	50,657	$100,411
70	Hudson, Ohio	22,382	$99,156
71	Bethesda, Maryland	55,300	$99,102
72	Greenwich, Connecticut	61,101	$99,086
73	Morganville, New Jersey	11,135	$99,035
74	Western Springs, Illinois	12,621	$98,876
75	Livingston, New Jersey	27,391	$98,869
76	Granite Bay, California	19,441	$98,762
77	Westfield, New Jersey	29,644	$98,390
78	Lower Makefield Township, Pennsylvania	32,681	$98,090
79	Moraga, California	16,642	$98,080
80	Union Hill-Novelty Hill, Washington	11,179	$98,061
81	Mendota Heights, Minnesota	11,071	$97,701
82	Thousand Oaks, California (includes Newbury Park, California)	117,005	$97,372
83	Coppell, Texas	35,955	$96,935
84	Lexington, MA	30,355	$96,825
85	Springdale, New Jersey	14,412	$96,412
86	Tustin Foothills, California	24,018	$96,230
87	San Ramon, California	44,477	$95,856
88	Northbrook, Illinois	33,425	$95,665
89	Rancho Palos Verdes, California	41,301	$95,503
90	Flower Mound, Texas	51,203	$95,416
91	Readington Township, New Jersey	15,803	$95,356
92	Foster City, California	28,803	$95,279
93	Olney, Maryland	31,651	$94,818
94	Germantown, Tennessee	37,281	$94,609
95	Palm Beach, Florida	10,374	$94,562
96	Los Gatos, California	28,683	$94,319
97	Winchester, Massachusetts	20,810	$94,049
98	Wildwood, Missouri	33,445	$94,006
99	New Territory, Texas	13,886	$93,972

Highest-income places with a population of at least 50,000

Rank	Place	Population	Median Household Income
1	Arcadia, California	56,364	$107,877
2	Bethesda, Maryland	55,300	$99,102
3	Greenwich, Connecticut	61,101	$99,086
4	Flower Mound, Texas	64,669	$95,416
5	Burke, Virginia	57,651	$93,561
6	West Bloomfield Township, Michigan	64,804	$91,661
7	Pleasanton, California	63,569	$90,859
8	Palo Alto, California	58,783	$90,377
9	Yorba Linda, California	58,595	$89,593
10	Naperville, Illinois	128,300	$88,771
11	Highlands Ranch, Colorado	70,864	$86,792
12	Newton, Massachusetts	83,829	$86,052
13	The Woodlands, Texas	55,824	$85,253
14	Milpitas, California	62,714	$84,429
15	Wayne, New Jersey	54,115	$83,651
17	Fairfield, Connecticut	57,340	$83,512
18	Newport Beach, California	70,022	$83,455
19	Sugar Land, Texas	63,507	$81,767
20	Laguna Niguel, California	61,963	$80,733
21	Reston, Virginia	56,447	$80,018
22	Ellicott City, Maryland	56,309	$79,031
23	Plano, Texas	222,301	$78,722
24	Eden Prairie, Minnesota	60,797	$78,328
25	Mission Viejo, California	92,780	$78,248
26	Troy, Michigan	80,959	$77,538
27	Plymouth, Minnesota	70,576	$77,008
28	Thousand Oaks, California (includes Newbury Park, California)	117,005	$76,815
29	Fremont, California	203,413	$76,579
30	Maple Grove, Minnesota	61,567	$76,111
31	Livermore, California	73,436	$75,322
32	Cary, North Carolina	94,530	$75,122
33	Rochester Hills, Michigan	68,840	$74,912
34	Sunnyvale, California	131,905	$74,409
35	Wheaton, Illinois	55,439	$73,385
36	Folsom, California	51,912	$73,175
37	Cerritos, California	51,507	$73,030
38	Annandale, Virginia	55,041	$72,561

39	Canton, Michigan	76,310	$72,495
40	Missouri City, Texas	52,477	$72,434
41	Irvine, California	212,793	$72,057
42	Union City, California	66,861	$71,926
43	Roswell, Georgia	79,844	$71,726
44	Columbia, Maryland	88,391	$71,524
45	Simi Valley, California	111,547	$70,370
46	San Jose, California	893,889	$70,243
47	Minnetonka, Minnesota	51,299	$69,979
48	Levittown, New York	53,063	$69,923
49	Edison, New Jersey	97,687	$69,746
50	Fountain Valley, California	54,995	$69,734
51	Santa Clara, California	102,104	$69,466
52	Mountain View, California	70,467	$69,362
53	Redondo Beach, California	63,261	$69,173
54	Diamond Bar, California	56,349	$68,871
55	Gilbert, Arizona	109,936	$68,032
56	Lake Forest, California	58,806	$67,967
57	Bolingbrook, Illinois	56,454	$67,852
58	Arlington Heights, Illinois	76,098	$67,807
59	Orland Park, Illinois	51,103	$67,574
60	Farmington Hills, Michigan	82,111	$67,493
61	Eagan, Minnesota	63,629	$67,388
62	Redwood City, California	75,447	$66,748
63	Santa Clarita, California	151,381	$66,717
64	Brookline, Massachusetts	57,061	$66,711
65	Sandy, Utah	88,259	$66,458
66	Hoffman Estates, Illinois	50,352	$65,937
67	Catalina Foothills, Arizona	53,560	$65,657
68	Dale City, Virginia	56,053	$65,355
69	Shelby Charter Township, Michigan	65,070	$65,291
70	Carlsbad, California	77,998	$65,145
71	Castro Valley, California	57,410	$64,874
72	Huntington Beach, California	189,940	$64,824
73	San Mateo, California	92,372	$64,757
74	Encinitas, California	58,195	$63,954
75	McKinney, Texas	54,384	$63,366
76	Aspen Hill, Maryland	50,223	$63,340
77	Palatine, Illinois	65,156	$63,321

78	Walnut Creek, California	64,583	$63,238
79	Livonia, Michigan	100,545	$63,018
80	Arlington, Virginia	189,453	$63,001
81	Tracy, California	56,839	$62,794
82	Camarillo, California	57,122	$62,457
83	Germantown, Maryland	55,152	$62,431
84	Carrollton, Texas	109,215	$62,406
85	Richardson, Texas	91,635	$62,392
86	Bellevue, Washington	109,189	$62,338
87	Daly City, California	103,549	$62,310
88	Overland Park, Kansas	148,848	$62,116
89	Hoover, Alabama	62,480	$61,982
90	South San Francisco, California	60,727	$61,764
91	Petaluma, California	54,538	$61,679
92	West Hartford, Connecticut	63,589	$61,665
93	Milford, Connecticut	50,602	$61,167
94	Olathe, Kansas	93,013	$61,111
95	San Rafael, California	56,132	$60,994
96	Schaumburg, Illinois	74,511	$60,941
97	Rancho Cucamonga, California	128,161	$60,931
98	Lee's Summit, Missouri	71,074	$60,905
99	Elk Grove, California	60,255	$60,661
100	Stamford, Connecticut	117,083	$60,556

Highest-income counties and places by state

For more detailed lists of rankings of counties and places in individual states, see the following pages:

- Alabama
- Alaska
- Arizona
- Arkansas
- California
- Colorado
- Connecticut
- Delaware
- Florida
- Georgia
- Hawaii
- Idaho
- Illinois
- Indiana
- Iowa

- Kansas
- Kentucky
- Louisiana
- Maine
- Maryland
- Massachusetts
- Michigan
- Minnesota
- Mississippi
- Missouri
- Montana
- Nebraska
- Nevada
- New Hampshire
- New Jersey
- New Mexico
- New York
- North Carolina
- North Dakota
- Oklahoma
- Ohio
- Oregon
- Pennsylvania
- Rhode Island
- South Carolina
- South Dakota
- Tennessee
- Texas
- Utah
- Vermont
- Virginia
- Washington
- West Virginia
- Wisconsin
- Wyoming

References

[1] Census 2000 Demographic Profiles (http://censtats.census.gov/pub/Profiles.shtml)

External links

- The Elite 100: America's Wealthiest Neighborhoods (http://higley1000.com/archives/44)
- Map of Cities by Median Income (http://www.easymapmaker.com/map/cities_median_income)

Poverty_threshold

The **poverty threshold**, or **poverty line**, is the minimum level of income deemed adequate in a given country.[1] In practice, like the definition of poverty, the official or common understanding of the poverty line is significantly higher in developed countries than in developing countries.[2] [3] The common **international poverty line** has in the past been roughly $1 a day.[4] In 2008, the World Bank came out with a revised figure of $1.25 at 2005 purchasing-power parity (PPP).[5]

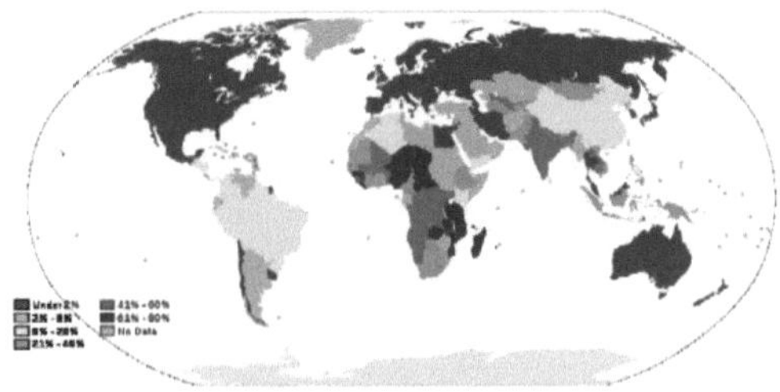

Percentage of population living on less than $1.25 per day. UN estimates 2000-2007.

Determining the poverty line is usually done by finding the total cost of all the essential resources that an average human adult consumes in one year.[6] The largest of these expenses is typically the rent required to live in an apartment, so historically, economists have paid particular attention to the real estate market and housing prices as a strong poverty line affector. Individual factors are often used to account for various circumstances, such as whether one is a parent, elderly, a child, married, etc. The poverty threshold may be adjusted annually.

History

Charles Booth, a pioneering investigator of poverty in London at the turn of the 20th century, popularised the idea of a *poverty line*, a concept originally conceived by the London School Board.[7] Booth set the line at 10 to 20 shillings, which he considered to be the minimum amount necessary for a family of 4 or 5 people to subsist.[8] Benjamin Seebohm Rowntree (1871–1954), a British sociological researcher, social reformer and industrialist, surveyed rich families in York and drew a poverty line in terms of a minimum weekly sum of money "necessary to enable families ... to secure the necessaries of a healthy life", which included fuel and light, rent, food, clothing, and household and personal items. Based on data from leading nutritionists of the period he calculated the cheapest price for the minimum calorific intake and nutritional balance necessary before people got ill or lost weight. He considered this amount to set his poverty line and concluded that 27.84 percent of the total population of York lived below this poverty line.[9] This result corresponded with that from Charles Booth's study of poverty in London and so challenged the view, commonly held at the time, that abject poverty was a problem particular to London and was not widespread in the rest of Britain. Rowntree distinguished between primary poverty, those who lacking in income and secondary poverty, those who had enough income but spent it elsewhere (1901:295-6).[9]

Absolute poverty

Absolute poverty is a level of poverty as defined in terms of the minimal requirements necessary to afford minimal standards of food, clothing, health care and shelter.[10] For the measure to be absolute, the line must be the same in different countries, cultures, and technological levels. Such an absolute measure should look only at the individual's power to consume and it should be independent of any changes in income distribution. The intuition behind an absolute measure is that mere survival takes essentially the same amount of resources across the world and that everybody should be subject to the same standards if meaningful comparisons of policies and progress are to be made. Notice that if everyone's real income in an economy increases, and the income distribution does not change, absolute poverty will decline.

Measuring poverty by an absolute threshold has the advantage of applying the same standard across different locations and time periods: it makes comparisons easier. On the other hand, it suffers from the disadvantage that any absolute poverty threshold is to some extent arbitrary; the amount of wealth required for survival is not the same in all places and time periods. For example, a person living in far northern Scandinavia requires a source of heat during colder months, while a person living on a tropical island does not.

This type of measure is often contrasted with measures of relative poverty (see below), which classify individuals or families as "poor" not by comparing them to a fixed cutoff point, but by comparing them to others in the population under study.

The term *absolute poverty* is also sometimes used as a synonym for extreme poverty. Absolute poverty is the absence of enough resources (such as money) to secure basic life necessities.

According to a UN declaration that resulted from the World Summit on Social Development in Copenhagen in 1995, absolute poverty is "a condition characterised by severe deprivation of basic human needs, including food, safe drinking water, sanitation facilities, health, shelter, education and information. It depends not only on income but also on access to services."[]

David Gordon's paper, "Indicators of Poverty & Hunger", for the United Nations, further defines absolute poverty as the absence of any two of the following eight basic needs:[]

- *Food:* Body Mass Index must be above 16.
- *Safe drinking water:* Water must not come solely from rivers and ponds, and must be available nearby (less than 15 minutes' walk each way).
- *Sanitation facilities:* Toilets or latrines must be accessible in or near the home.
- *Health:* Treatment must be received for serious illnesses and pregnancy.
- *Shelter:* Homes must have fewer than four people living in each room. Floors must not be made of dirt, mud, or clay.
- *Education:* Everyone must attend school or otherwise learn to read.
- *Information:* Everyone must have access to newspapers, radios, televisions, computers, or telephones at home.
- *Access to services:* This item is undefined by Gordon, but normally is used to indicate the complete panoply of education, health, legal, social, and financial (credit) services.

For example, a person who lives in a home with a mud floor is considered severely deprived of shelter. A person who never attended school and cannot read is considered severely deprived of education. A person who has no newspaper, radio, television, or telephone is considered severely deprived of information. All people who meet any two of these conditions — for example, they live in homes with mud floors *and* cannot read — are considered to be living in absolute poverty.

Relative poverty

Relative poverty is the "most useful measure for ascertaining poverty rates in wealthy developed nations".[] [11] [] [12] [] Relative poverty measure is used by the United Nations Development Program (UNDP), the United Nations Children's Fund (UNICEF), the Organisation for Economic Co-operation and Development (OECD) and Canadian poverty researchers.[] [11] [] [12] [] In the European Union, the "relative poverty measure is the most prominent and most–quoted of the EU social inclusion indicators."[]

"[R]elative poverty reflects better the cost of social inclusion and equality of opportunity in a specific time and space."[13]

"Once economic development has progressed beyond a certain minimum level, the rub of the poverty problem – from the point of view of both the poor individual and of the societies in which they live – is not so much the effects of poverty in any absolute form but the effects of the contrast, daily perceived, between the lives of the poor and the lives of those around them. For practical purposes, the problem of poverty in the industrialized nations today is a problem of relative poverty (page 9)."[14] [13]

History of the concept of relative poverty

In 1776 Adam Smith in the Wealth of Nations argued that poverty is the inability to afford, "not only the commodities which are indispensably necessary for the support of life, but whatever the custom of the country renders it indecent for creditable people, even of the lowest order, to be without."[] [15]

In 1958 J. K. Galbraith argued that, "People are poverty stricken when their income, even if adequate for survival, falls markedly behind that of their community."[] [15]

In 1964 in a joint committee economic President's report in the United States, Republicans endorsed the concept of relative poverty. "No objective definition of poverty exists.... The definition varies from place to place and time to time. In America as our standard of living rises, so does our idea of what is substandard."[16] [15]

In 1965 Rose Friedman argued for the use of relative poverty claiming that the definition of poverty changes with general living standards. Those labelled as poor in 1995, would have had "a higher standard of living than many labelled not poor" in 1965.[17] [15]

In 1979, British sociologist, Peter Townsend published his famous definition, "individuals [...] can be said to be in poverty when they lack the resources to obtain the types of diet, participate in the activities and have the living conditions and amenities which are customary, or are at least widely encouraged or approved, in the societies to which they belong (page 31)."[]

Brian Nolan and Christopher T. Whelan of the Economic and Social Research Institute (ESRI) in Ireland explained that, "[P]overty has to be seen in terms of the standard of living of the society in question."[]

Relative poverty measures are used as official poverty rates by the European Union, UNICEF and the OEDC. The main poverty line used in the OECD and the European Union is based on "economic distance", a level of income set at 60% of the median household income.[18]

Relative poverty versus inequality

A measure of *relative poverty* defines "poverty" as being below some relative poverty threshold. For example, the statement that "those individuals who are employed and whose household equivalised disposable income is below 60% of national median equivalised income are poor" uses a relative measure to define poverty.[] Using this definition, if everyone's real income in an economy increases, but the income distribution stays the same, then the rate of relative poverty will also stay the same.

This means, by its very nature, that there will always be a family living in (relative) poverty, even if they have a very high living standard, unless everyone has almost exactly the same income.

Relative poverty measurements can sometimes produce odd results, especially in small populations. For example, if the median household in a wealthy neighborhood earns US$1 million each year, then a family that earns US$100,000 would be considered poor on the relative poverty scale, even though such a family could meet all of its basic needs and much more. At the other end of the scale, if the median household in a very poor neighborhood earned only 50% of what it needs to buy food, then a person who earned the median income would not be considered poor on a relative poverty scale, even though the person is clearly poor on an absolute poverty scale.

Measures of relative poverty are almost the same as measuring income inequality: If a society gets a more equal income distribution, relative poverty will fall. They point out that if society changed in a way that hurt high earners more than low ones, then relative poverty would decrease, but every citizen of the society would be worse off. Likewise in the reverse direction: it is possible to reduce absolute poverty while increasing relative poverty.

The term *relative poverty* can also be used in a different sense to mean "moderate poverty" – for example, a standard of living or level of income that is high enough to satisfy basic needs (like water, food, clothing, shelter, and basic health care), but still significantly lower than that of the majority of the population under consideration[19] .

Basic needs

See article Basic needs

The basic needs approach is one of the major approaches to the measurement of absolute poverty in developing countries. It attempts to define the absolute minimum resources necessary for long-term physical well-being, usually in terms of consumption goods. The poverty line is then defined as the amount of income required to satisfy those needs. The 'basic needs' approach was introduced by the International Labour Organization's World Employment Conference in 1976.[] [] "Perhaps the high point of the WEP was the World Employment Conference of 1976, which proposed the satisfaction of basic human needs as the overriding objective of national and international development policy. The basic needs approach to development was endorsed by governments and workers' and employers' organizations from all over the world. It influenced the programmes and policies of major multilateral and bilateral development agencies, and was the precursor to the human development approach."[] []

A traditional list of immediate "basic needs" is food (including water), shelter and clothing.[] Many modern lists emphasize the minimum level of consumption of 'basic needs' of not just food, water, and shelter, but also sanitation, education, and health care. Different agencies use different lists.

In 1978 Ghai investigated the literature that criticized the basic needs approach. Critics argued that the basic needs approach lacked scientific rigour; it was consumption-oriented and anti-growth. Some considered it to be "a recipe for perpetuating economic backwardness" and for giving the impression "that poverty elimination is all too easy".[] Amartya Sen focused on 'capabilities' rather than consumption.

In the development discourse, the basic needs model focuses on the measurement of what is believed to be an eradicable level of poverty.

Professor Chris Sarlo, an economist at Nipissing University in North Bay, Canada, and a senior fellow of the Fraser Institute, used Statistics Canada's socio-economic databases, particularly the *Survey of Household Spending* to determine the cost of a list of household necessities. The list includes food, shelter, clothing, health care, personal care, essential furnishings, transportation and communication, laundry, home insurance, and miscellaneous; it assumes that education is provided freely to all residents of Canada. This is calculated for various communities across Canada and adjusted for family size. With this information, he determines the proportion of Canadian households that have insufficient income to afford those necessities. Sarlo introduced his basic needs approach in 1992 in his book entitled *Poverty in Canada* published by the Fraser Institute. Sarlos argued that Statistics Canada's Low Income Cutoff (LICO), which uses a relative poverty line threshold was inadequate. Sarlos' basic needs poverty measure for Canada was updated in 2001 Sarlos' to reflect "new developments in the literature, external critiques of the approach, and better (and more comprehensive) data". For example it included out-of-pocket health-care costs. According to the Sarlo, "the basic-needs approach is partly absolute (the list [of necessities] is limited to items required for long-term physical well-being) and partly relative, reflecting the standards that apply in the individual's own society at the present time." It was updated again in 2006. [] The Fraser Institute's list of necessities for living creditably in Canada includes not only food, shelter, clothing, and health care, but also personal care, furniture, transportation, communication, laundry, and home insurance. It is criticized for not including any entertainment items like cable television, daily newspapers, and tickets to movies or sporting events.[] [] [20] [20] Based on his basic needs poverty threshold, the poverty rate in Canada, the poverty rate has declined from about 12% of Canadian households to about 5% since the 1970s.[21] This is in sharp contrast to Statistic Canada, the Conference Board of Canada, the Organisation for Economic Co-operation and Development (OECD) and UNESCO reports on poverty in Canada. They used the relative poverty measure considered to the most useful for advanced industrial nations like Canada, which Sarlo rejects.

According to a 2008 report by the Organisation for Economic Co-operation and Development (OECD), the rate of poverty in Canada, is among the highest of the OECD member nations, the world's wealthiest industrialized nations.[] There is no official government definition and therefore, measure, for poverty in Canada. However, Raphael Dennis, author of *Poverty in Canada: Implications for Health and Quality of Life*[] [] reported that the

United Nations Development Program (UNDP), the United Nations Children's Fund (UNICEF), the Organisation for Economic Co-operation and Development (OECD) and Canadian poverty researchers[22] [] find that relative poverty is the "most useful measure for ascertaining poverty rates in wealthy developed nations such as Canada".[] [11] [] [12] []

National poverty lines

National estimates are based on population-weighted subgroup estimates from household surveys. Definitions of the poverty line may vary considerably among nations. For example, rich nations generally employ more generous standards of poverty than poor nations. Thus, the numbers are not comparable among countries.

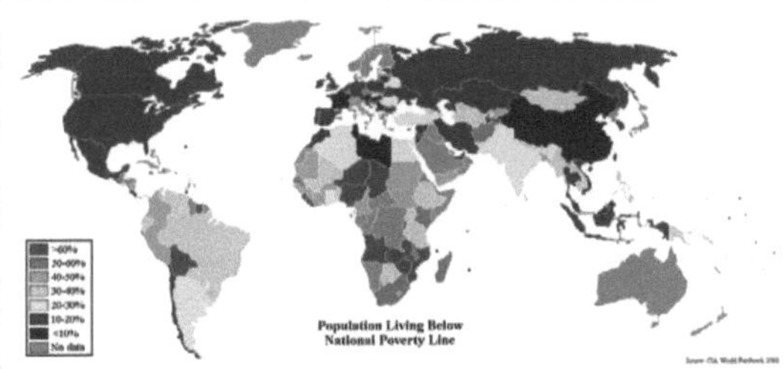

CIA World Factbook-based map showing the percentage of population by country living below that country's official poverty line.

In 2010, in the United States, the poverty threshold for one person under 65 was US$11,344 (annual income); the threshold for a family group of four, including two children, was US$22,133.[] [23] According to the U.S. Census Bureau data released on September 13, 2011, the nation's poverty rate rose to 15.1 percent in 2010.

In the UK, "more than five million people – over a fifth (23 percent) of all employees – were paid less than £6.67 an hour in April 2006. This is based on a low pay rate of 60 percent of full-time median earnings, equivalent to a little over £12,000 a year for a 35-hour working week. In April 2006, a 35-hour week would have earned someone £9,191 a year – before tax or National Insurance".[24] [25]

India's official poverty level, on the other hand, is split according to rural vs. urban thresholds. For urban dwellers, the poverty line is defined as living on less than 538.60 rupees (approximately USD $12) per month, whereas for rural dwellers, it is defined as living on less than 356.35 rupees per month (approximately USD $7.50).[]

Criticisms

Using a poverty threshold is problematic because having an income marginally above it is not substantially different from having an income marginally below it: the negative effects of poverty tend to be continuous rather than discrete, and the same low income affects different people in different ways. To overcome this problem, poverty indices are sometimes used instead; see income inequality metrics.

A poverty threshold relies on a quantitative, or purely numbers-based, measure of income. If other human development-indicators like health and education are used, they must be quantified, which is not a simple (if even achievable) task.

Overstating poverty

In-kind transfers, whether from public or private sources, are not counted when calculating a poverty threshold. For example, if a parent pays the rent on an apartment for an adult son directly to the apartment owner, instead of giving the money to the son to pay the rent, then that money does not count as income to the son. If a church or non-profit organization gives food to an elderly person, the value of the food is not counted as income to the elderly person. Rea Hederman, a senior policy analyst in the Center for Data Analysis at the Heritage Foundation, said

> The official poverty measure counts only monetary income. It considers antipoverty programs such as food stamps, housing assistance, the Earned Income Tax Credit, Medicaid and school lunches, among

others, "in-kind benefits" – and hence not income. So, despite everything these programs do to relieve poverty, they aren't counted as income when Washington measures the *poverty rate.*[26]

Studies measuring the difference between income before and after taxes and government transfers, however, have found that without these programs poverty would be roughly 30% to 40% higher than the official poverty line indicates, despite many of their benefits not being counted as income.[27] [28]

Further, the U.S. Census Bureau calculates the poverty line the same throughout the U.S. regardless of the cost-of-living in a state or urban area. For instance, the cost-of-living in California, the most populous state, was 1.42 times the U.S. average in 2010 while the cost-of-living in Texas, the second most populous state, was 0.90 (90%) of the U.S. average.

Notes

[1] Ravallion, Martin *Poverty freak: A Guide to Concepts and Methods*. Living Standards Measurement Papers, The World Bank, 1992, p. 25

[2] Hagenaars, Aldi & de Vos, Klaas *The Definition and Measurement of Poverty*. Journal of Human Resources, 1988

[3] Hagenaars, Aldi & van Praag, Bernard *A Synthesis of Poverty Line Definitions*. Review of Income and Wealth, 1985

[4] Sachs, Jeffrey D. *The End of Poverty* 2005, p. 20

[5] Ravallion, Martin; Chen Shaohua & Sangraula, Prem *Dollar a day* The World Bank Economic Review, 23, 2, 2009, pp. 163-184

[6] {Poverty Lines-Martin Ravallion, in The New Palgrave Dictionary of Economics, 2nd Edition, London: Palgrave Macmillan}

[7] Alan Gillie, 'The Origin of the Poverty Line', Economic History Review, XLIX/4 (1996), 726

[8] David Boyle - The Tyranny of Numbers p.116

[9] page 298

[11] Child poverty in rich nations: Report card no. 6 (Report). Innocenti Research Centre. 2005.

[12] Human development report: Capacity development: Empowering people and institutions (Report). Geneva: United Nations Development Program. 2008.

[13] Jonathan Bradshaw; Yekaterina Chzhen; Gill Main; Bruno Martorano; Leonardo Menchini; Chris de Neubourg (January 2012) (PDF). [www.unicef-irc.org/publications/pdf/iwp_2012_01.pdf Relative Income Poverty among Children in Rich Countries] (Report). Innocenti Working Paper. Florence, Italy: UNICEF Innocenti Research Centre. ISSN 1014-7837. www.unicef-irc.org/publications/pdf/iwp_2012_01.pdf.

[14] A League Table of Child Poverty in Rich Nations (Report). Innocenti Report Card No.1. Florence, Italy: UNICEF Innocenti Research Centre.

[15] Peter Adamson (2012) (PDF). Measuring child poverty: New league tables of child poverty in the world's rich countries (http://www.unicef.ca/sites/default/files/imce_uploads/DISCOVER/OUR WORK/ADVOCACY/DOMESTIC/POLICY ADVOCACY/DOCS/unicefreportcard10-eng.pdf) (Report). UNICEF Innocenti Research Centre Report Card. Florence, Italy. .

[16] Minority [Republican] views, p. 46 in U.S. Congress, Report of the Joint Economic Committee on the January 1964 Economic Report of the President with Minority and Additional Views (Report). Washington, D.C.: US Government Printing Office. January 1964.

[17] Friedman, Rose. D. (1965). " *Poverty: Definition and Perspective* ". *American Enterprise Institute for Public Policy Research* (Washington, D.C.).

[19] http://www.caledonia.org.uk/papers/Who-and-Where-are-the-Poor.pdf

[20] Chris Sarlo (23 July 2001). Measuring Poverty in Canada (Report). Fraser Institute.

[21] Poverty in Canada: 2006 Update (http://www.fraserinstitute.org/COMMERCE.WEB/product_files/PovertyinCanada2006.pdf)

[22] The Conference Board of Canada "uses the OECD's relative measure of child poverty, which calculates the proportion of children living in households where disposable income is less than 50 per cent of the median in each country." The Conference Board 2013 cautioned that Canada's high poverty rate, ranks among the worst of the 17 countries they compared. "Canada's child poverty rate was 15.1 per cent, up from 12.8 per cent in the mid-1990s. Only the United States ranked lower.

[24] Working out of Poverty: A study of the low paid and the working poor by Graeme Cooke and Kayte Lawton

[25] IPPR Article: "Government must rescue 'forgotten million children' in poverty" (http://www.ippr.org.uk/pressreleases/?id=2965)

[26] "Poor poverty yardsticks" (http://www.heritage.org/Press/Commentary/ed090806b.cfm), Rea S. Hederman, Jr., September 8, 2006, The Heritage Foundation

[27] Kenworthy, L. (1999). Do social-welfare policies reduce poverty? A cross-national assessment. *Social Forces, 77*(3), 1119-1139. doi:10.1093/sf/77.3.1119

[28] Bradley, D., Huber, E., Moller, S., Nielson, F. & Stephens, J. D. (2003). Determinants of relative poverty in advanced capitalist democracies. (http://www.jstor.org/stable/3088901) *American Sociological Review, 68*(3), 22-51.

See also

- Asset poverty
- Income deficit
- List of countries by percentage of population living in poverty
- Living wage
- Minimum wage
- Measuring poverty
- UN Millennium Development Goals

References

Further reading

- Alan Gillie, "The Origin of the Poverty Line", *Economic History Review,* XLIX/4 (1996), 726

External links

- History of the U.S. Poverty Line (http://eesc.orst.edu/agcomwebfile/edmat/html/em/em8743/part1/officialgovt.html) by Tom Gentle, Oregon State University.
- United States Department of Health and Human Services Poverty Guidelines, Research, and Measurement (http://aspe.hhs.gov/poverty/index.shtml)
- 2007 United States Department of Health and Human Services Poverty Guidelines (http://aspe.hhs.gov/poverty/07poverty.shtml)
- Debraj Ray 1998, *Development Economics*, Princeton University Press, ISBN 0-691-01706-9.
- World Summit for Social Development Agreements (http://www.un.org/esa/socdev/wssd/agreements/index.html), United Nations
- The "elimination" of poverty (http://www.inclusivedemocracy.org/journal/vol4/vol4_no1_takis_poverty.htm), Takis Fotopoulos, *The International Journal of Inclusive Democracy* (http://www.inclusivedemocracy.org/journal/), vol.4, no.1 (January 2008).

Time_zone

A **time zone** is a region on Earth that has a uniform standard time for legal, commercial, and social purposes. It is convenient for areas in close commercial or other communication to keep the same time, so time zones tend to follow the boundaries of countries and their subdivisions.

Most of the time zones on land are offset from Coordinated Universal Time (UTC) by a whole number of hours (UTC−12 to UTC+14), but a few are offset by 30 or 45 minutes. Some higher latitude countries use daylight saving time for part of the year, typically by changing clocks by an hour. Many land time zones are skewed toward the west of the corresponding nautical time zones. This also creates a permanent daylight saving time effect.

History

Early timekeeping

Before the invention of clocks, people marked the time of day with apparent solar time (also called "true" solar time) – for example, the time on a sundial – which was typically different for every settlement.

When well-regulated mechanical clocks became widespread in the early 19th century, each city began to use some local mean solar time. Apparent and mean solar time can differ by up to around 15 minutes (as described by the equation of time) due to the non-circular shape of the Earth's orbit around the sun and the tilt of the Earth's axis. Mean solar time has days of equal length, and the difference between the two averages to zero after a year.

Greenwich Mean Time (GMT) was established in 1675 when the Royal Observatory was built as an aid to mariners to determine longitude at sea, providing a standard reference time when each city in England kept a different local time.

Railroad time

Local solar time became increasingly awkward as railways and telecommunications improved, because clocks differed between places by an amount corresponding to the difference in their geographical longitude, which varied by four minutes for every degree of longitude. The difference between New York and Boston is about two degrees or 8 minutes, the difference between Sydney and Melbourne, Australia, is about 7 degrees or 28 minutes. Bristol is 2°35′ W(est) of Greenwich (East London), so when it is noon in Bristol, it is about 10 minutes past noon in London.[1] The use of time zones smooths out these differences.

Plaque commemorating the Railway General Time Convention of 1883 in North America

The first adaptation of a standard time in the world was established on December 1, 1847, in Great Britain by railway companies using GMT kept by portable chronometers. The first of these companies to adopt standard time was the Great Western Railway (GWR) in November 1840. This quickly became known as Railway Time. About August 23, 1852, time signals were first transmitted by telegraph from the Royal Observatory, Greenwich. Even though 98% of Great Britain's public clocks were using GMT by 1855, it was not made Britain's legal time until August 2, 1880. Some old British clocks from this period have two minute hands—one for the local time, one for GMT.[2]

The increase in worldwide communication had further increased the need for interacting parties to communicate mutually comprehensible time references to one another. The problem of differing local times could be solved across larger areas by synchronizing clocks worldwide, but in many places the local time would then differ markedly from the solar time to which people were accustomed. Time zones were a compromise, relaxing the complex geographic dependence while still allowing local time to approximate the mean solar time.

On November 2, 1868, the then-British colony of New Zealand officially adopted a standard time to be observed throughout the colony, and was perhaps the first country to do so. It was based on the longitude 172°30′ East of Greenwich, that is 11 hours 30 minutes ahead of GMT. This standard was known as New Zealand Mean Time.

Timekeeping on the American railroads in the mid-19th century was somewhat confused. Each railroad used its own standard time, usually based on the local time of its headquarters or most important terminus, and the railroad's train schedules were published using its own time. Some major railroad junctions served by several different railroads had a separate clock for each railroad, each showing a different time; the main station in Pittsburgh, Pennsylvania, for example, kept six different times.

Charles F. Dowd proposed a system of one-hour standard time zones for American railroads about 1863, although he published nothing on the matter at that time and did not consult railroad officials until 1869. In 1870, he proposed four ideal time zones (having north–south borders), the first centered on Washington, D.C., but by 1872 the first was centered 75°W of Greenwich, with geographic borders (for example, sections of the Appalachian Mountains). Dowd's system was never accepted by American railroads. Instead, U.S. and Canadian railroads implemented a version proposed by William F. Allen, the editor of the *Traveler's Official Railway Guide*.[3] The borders of its time zones ran through railroad stations, often in major cities. For example, the border between its Eastern and Central time zones ran through Detroit, Buffalo, Pittsburgh, Atlanta, and Charleston. It was inaugurated on Sunday, November 18, 1883, also called "The Day of Two Noons",[4] when each railroad station clock was reset as standard-time noon was reached within each time zone. The zones were named Intercolonial, Eastern, Central, Mountain, and Pacific. Within one year, 85% of all cities with populations over 10,000, about 200 cities, were using standard time.[5] A notable exception was Detroit (which is about half-way between the meridians of eastern time and central time), which kept local time until 1900, then tried Central Standard Time, local mean time, and Eastern Standard Time before a May 1915 ordinance settled on EST and was ratified by popular vote in August 1916. The confusion of times came to an end when Standard zone time was formally adopted by the U.S. Congress in the Standard Time Act of March 19, 1918.

Worldwide time zones

Although the first person to propose a worldwide system of time zones was Italian mathematician Quirico Filopanti in his book *Miranda!* published in 1858, his idea was unknown outside the pages of his book until long after his death, so it did not influence the adoption of time zones during the 19th century. He proposed 24 hourly time zones, which he called "longitudinal days", the first centered on the meridian of Rome. He also proposed a universal time to be used in astronomy and telegraphy.[6] [7]

Scottish-born Canadian Sir Sandford Fleming proposed a worldwide system of time zones in 1879. He advocated his system at several international conferences, thus is widely credited with their invention. In 1876, his first proposal was for a global 24-hour clock, conceptually located at the center of the Earth and not linked to any surface meridian. In 1879 he specified that his universal day would begin at the anti-meridian of Greenwich (180th meridian), while conceding that hourly time zones might have some limited local use. He also proposed his system at the International Meridian Conference in October 1884, but it did not adopt his time zones because they were not within its purview. The conference did adopt a universal day of 24 hours beginning at Greenwich midnight, but specified that it "shall not interfere with the use of local or standard time where desirable".

By about 1900, almost all time on Earth was in the form of standard time zones, only some of which used an hourly offset from GMT. Many applied the time at a local astronomical observatory to an entire country, without any reference to GMT. It took many decades before all time on Earth was in the form of time zones referred to some "standard offset" from GMT/UTC. Most major countries had adopted hourly time zones by 1929. Nepal was the last country to adopt a standard offset, shifting slightly to UTC+5:45 in 1986.

Today, all nations use standard time zones for secular purposes, but they do not all apply the concept as originally conceived. Newfoundland, India, Iran, Afghanistan, Venezuela, Burma, the Marquesas, as well as parts of Australia use half-hour deviations from standard time, and some nations, such as Nepal, and some provinces, such as the Chatham Islands, use quarter-hour deviations. Some countries, most notably China and India, use a single time zone, even though the extent of their territory far exceeds 15° of longitude. Before 1949, China used five time zones (see Time in China).

Definition

Before 1972, all time zones were specified as an offset from Greenwich Mean Time (GMT), which was the mean solar time at the meridian passing through the Royal Observatory in Greenwich, London. Since 1972 all official time services have broadcast radio time signals synchronized to UTC, a form of atomic time that includes leap seconds to keep it within 0.9 seconds of this former GMT, now called UT1. Many countries now legally define their standard time relative to UTC, although some still legally refer to GMT, including the United Kingdom itself. UTC, also called Zulu time, is used everywhere on Earth by astronomers and others who need to state the time of an event unambiguously.

Time zones are based on Greenwich Mean Time (GMT), the mean solar time at longitude 0° (the Prime Meridian). The definition of GMT was recently changed – it was previously the same as UT1, a mean solar time calculated directly from the rotation of the Earth. As the rate of rotation of the Earth is not constant, the time derived from atomic clocks was adjusted to closely match UT1. In January 1972, however, the length of the second in both Greenwich Mean Time and atomic time was equalized. The readings of participating atomic clocks are averaged out to give a uniform time scale.

Because the length of the average day is a small fraction of a second more than 24 hours (slightly more than 86400 seconds), leap seconds are periodically inserted into Greenwich Mean Time to make it approximate to UT1. This new time system is also called Coordinated Universal Time (UTC). Leap seconds are inserted to keep UTC within 0.9 seconds of UT1. Because the Earth's rotation is gradually slowing, leap seconds will need to be added more frequently in the future. However, from one year to the next the rotation rate is slightly irregular, so leap seconds are not added unless observations of Earth's rotation show that one is required. In this way, local times will continue to stay close to mean solar time and the effects of variations in Earth's rotation rate will be confined to simple step changes relative to the uniform time scale (International Atomic Time or TAI). All local times differ from TAI by an integral number of seconds. With the implementation of UTC, nations began to use it in the definition of their time zones. As of 2005, most nations had altered the definition of local time in this way.

In the UK, this involved redefining Greenwich Mean Time to make it the same as UTC.[8] British Summer Time (BST) is still one hour in advance of Greenwich Mean Time and is therefore also one hour in advance of Coordinated Universal Time. Thus Greenwich Mean Time is the local time at the Royal Observatory, Greenwich between 0100 hours GMT on the last Sunday in October and 0100 hours GMT on the last Sunday in March. Similar circumstances apply in many other places.

Leap seconds are considered by many to be a nuisance, and ways to abolish them are being considered. This means letting the time difference accumulate. One suggestion is to insert a "leap-hour" in about 5,000 years. For more on this discussion read Proposal to abolish leap seconds.

Notation of time

ISO 8601

UTC

If the time is in Coordinated Universal Time (UTC), add a "Z" directly after the time without a space. "Z" is the zone designator for the zero UTC offset. "09:30 UTC" is therefore represented as "09:30Z" or "0930Z". "14:45:15 UTC" would be "14:45:15Z" or "144515Z".

UTC time is also known as "Zulu" time, since "Zulu" is the ICAO spelling alphabet code word for "Z".

Offsets from UTC

Offsets from UTC are written in the format ±[hh]:[mm], ±[hh][mm], or ±[hh]. So if the time being described is one hour ahead of UTC (such as the time in Berlin during the winter), the zone designator would be "+01:00", "+0100", or simply "+01". This is appended to the time in the same way that 'Z' was above. The offset from UTC changes with daylight saving time, e.g. a time offset in Chicago, which is in the North American Central Time Zone, would be "−06:00" for the winter (Central Standard Time) and "−05:00" for the summer.

Abbreviations

Time zones are often represented by abbreviations such as "EST, WST, CST" but these are not part of the international time and date standard ISO 8601 and their use as sole designator for a time zone is not recommended. Such designations can be ambiguous. For example, "BST", which is British Summer Time, was renamed "British Standard Time" between 1968 and 1971 when Central European Time was in force because legislators objected to calling it Central European Time. The same legislation affirmed that the Standard Time within the United Kingdom was, and would continue to be, Greenwich Mean Time.

UTC offsets worldwide

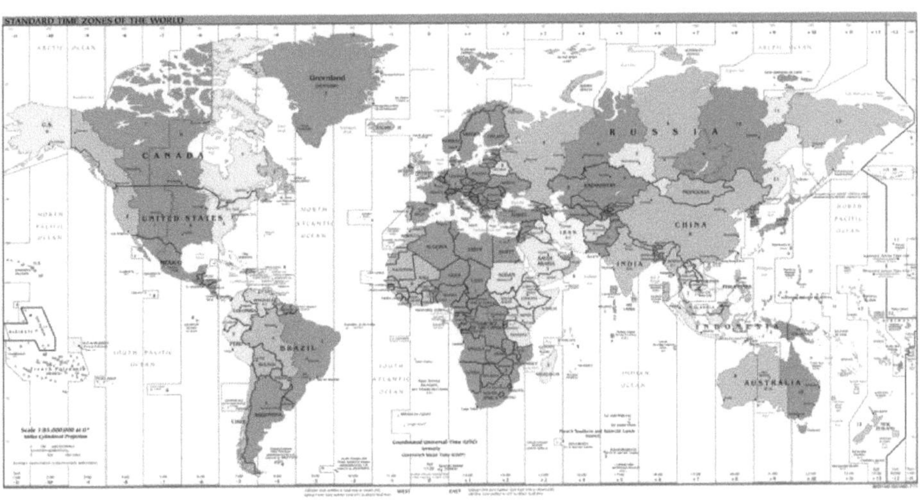

Time zones map (Last update: March 2013)

List of UTC offsets

These examples give the local time at various locations around the world when daylight saving time is not in effect:

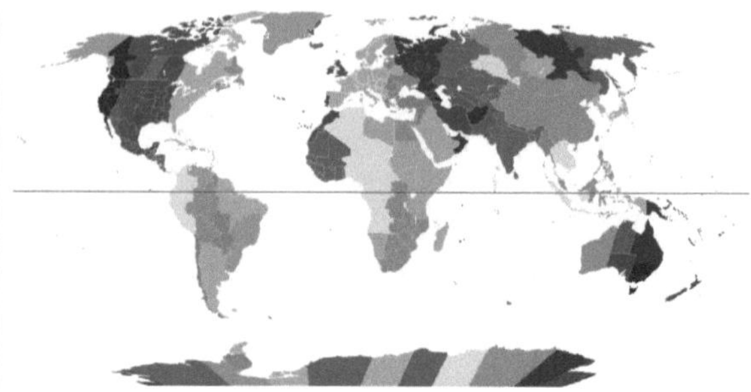

XX = ISO 3166-1 alpha-2 country code, XX- = parts of the country, N = North, S = South, UTC = Universal Coordinated Time, DST = Daylight Saving Time

Time zone	Example time (ISO 8601 notation)	Example locations	Example locations that in summer use DST
UTC−12:00	2013-08-13T04:24:33-12:00	Baker Island, Howland Island (both uninhabited)	
UTC−11:00	2013-08-13T05:24:33-11:00	American Samoa, Niue	
UTC−10:00	2013-08-13T06:24:33-10:00	United States (Hawaii)	Most of French Polynesia, United States (Aleutian Islands)
UTC−09:30	2013-08-13T06:54:33-09:30	Marquesas Islands	
UTC−09:00	2013-08-13T07:24:33-09:00	Gambier Islands	United States (most of Alaska)
UTC−08:00	2013-08-13T08:24:33-08:00		Canada (most of British Columbia), Mexico (Baja California), United States (California, most of Nevada, most of Oregon, Washington (state))
UTC−07:00	2013-08-13T09:24:33-07:00	Canada (northeastern British Columbia), Mexico (Sonora), United States (Arizona)	Canada (Alberta), Mexico (Chihuahua), United States (Colorado)
UTC−06:00	2013-08-13T10:24:33-06:00	Canada (almost all of Saskatchewan), Costa Rica, El Salvador, Ecuador (Galápagos Islands), Guatemala, Honduras, Mexico (most), Nicaragua,	Canada (Manitoba), United States (Illinois, most of Texas)
UTC−05:00	2013-08-13T11:24:33-05:00	Colombia, Cuba, Ecuador (continental), Jamaica, Panama, Peru	Canada (most of Ontario, most of Quebec), Haiti, United States (most of Florida, Georgia, Massachusetts, most of Michigan, most of Indiana, New York, North Carolina, Ohio, Washington, D.C.)
UTC−04:30	2013-08-13T11:54:33-04:30	Venezuela	
UTC−04:00	2013-08-13T12:24:33-04:00	Bolivia, Brazil (Amazonas), Chile (continental), Dominican Republic, Canada (Nova Scotia), Puerto Rico, Trinidad and Tobago	Falkland Islands
UTC−03:30	2013-08-13T12:54:33-03:30		Canada (island of Newfoundland and southern Labrador)
UTC−03:00	2013-08-13T13:24:33-03:00	Argentina, Paraguay	Brazil (Federal District, Rio de Janeiro, São Paulo), most of Greenland, Uruguay
UTC−02:00	2013-08-13T14:24:33-02:00	Brazil (Fernando de Noronha), South Georgia and the South Sandwich Islands	

UTC−01:00	2013-08-13T15:24:33-01:00	Portugal (Azores), Cape Verde	
UTC±00:00	2013-08-13T16:24:33+00:00	Côte d'Ivoire, Faroe Islands, Ghana, Iceland, Senegal	Ireland, Portugal (continental and Madeira), Spain (Canary Islands), Morocco, United Kingdom, Western Sahara
UTC+01:00	2013-08-13T17:24:33+01:00	Algeria, Angola, Benin, Cameroon, Gabon, Niger, Nigeria, Democratic Republic of the Congo (west), Tunisia	Albania, Austria, Belgium, Bosnia and Herzegovina, Spain (continental), Croatia, Czech Republic, Denmark, Germany, Hungary, Italy, Libya, Malta, Kosovo, Macedonia, France (metropolitan), Namibia, the Netherlands, Norway, Poland, Serbia, Slovakia, Slovenia, Sweden, Switzerland
UTC+02:00	2013-08-13T18:24:33+02:00	Egypt, Malawi, Mozambique, South Africa, Swaziland, Zambia, Zimbabwe	Bulgaria, Cyprus, Estonia, Finland, Greece, Israel, Latvia, Lebanon, Lithuania, Moldova, Northern Cyprus, Palestine, Romania, Syria, Transnistria, Turkey, Ukraine
UTC+03:00	2013-08-13T19:24:33+03:00	Belarus, Djibouti, Eritrea, Ethiopia, Iraq, Jordan, Kenya, Madagascar, Russia (Kaliningrad Oblast), Saudi Arabia, South Sudan, Sudan, Somalia, Somaliland, Tanzania, Uganda, Yemen	
UTC+03:30	2013-08-13T19:54:33+03:30		Iran
UTC+04:00	2013-08-13T20:24:33+04:00	Abkhazia, Armenia, Azerbaijan, Georgia, Mauritius, Oman, Russia (European), Seychelles, South Ossetia, United Arab Emirates	
UTC+04:30	2013-08-13T20:54:33+04:30	Afghanistan	
UTC+05:00	2013-08-13T21:24:33+05:00	Kazakhstan (west), Maldives, Pakistan, Uzbekistan	
UTC+05:30	2013-08-13T21:54:33+05:30	India, Sri Lanka	
UTC+05:45	2013-08-13T22:09:33+05:45	Nepal	
UTC+06:00	2013-08-13T22:24:33+06:00	Kazakhstan (most), Bangladesh, Bhutan, Russia (Ural: Sverdlovsk Oblast, Chelyabinsk Oblast)	
UTC+06:30	2013-08-13T22:54:33+06:30	Cocos Islands, Myanmar	
UTC+07:00	2013-08-13T23:24:33+07:00	Jakarta, Russia (Novosibirsk Oblast), Thailand, Vietnam	
UTC+08:00	2013-08-14T00:24:33+08:00	China, Hong Kong, Russia (Krasnoyarsk Krai), Malaysia, Philippines, Singapore, Taiwan, most of Mongolia, Western Australia	
UTC+08:45	2013-08-14T01:09:33+08:45	Australia (Eucla)	
UTC+09:00	2013-08-14T01:24:33+09:00	East Timor, Russia (Irkutsk Oblast), Japan, North Korea, South Korea	
UTC+09:30	2013-08-14T01:54:33+09:30	Australia (Northern Territory)	Australia (South Australia)
UTC+10:00	2013-08-14T02:24:33+10:00	Russia (Zabaykalsky Krai), Papua New Guinea, Australia (Queensland)	Australia (New South Wales, Tasmania, Victoria)
UTC+10:30	2013-08-14T02:54:33+10:30		Lord Howe Island
UTC+11:00	2013-08-14T03:24:33+11:00	New Caledonia, Russia (Primorsky Krai), Solomon Islands	
UTC+11:30	2013-08-14T03:54:33+11:30	Norfolk Island	

UTC+12:00	2013-08-14T04:24:33+12:00	Kiribati (Gilbert Islands), Fiji, Russia (Kamchatka Krai)	New Zealand
UTC+12:45	2013-08-14T05:09:33+12:45		Chatham Islands
UTC+13:00	2013-08-14T05:24:33+13:00	Kiribati (Phoenix Islands), Tonga, Tokelau	Samoa
UTC+14:00	2013-08-14T06:24:33+14:00	Kiribati (Line Islands)	

Where the adjustment for time zones results in a time at the other side of midnight from UTC, then the date at the location is one day later or earlier.

Some examples when UTC is 23:00 on Monday when or where daylight saving time is not in effect:

- Cairo, Egypt: UTC+02; 01:00 on Tuesday
- Wellington, New Zealand: UTC+12; 11:00 on Tuesday

Some examples when UTC is 02:00 on Tuesday when or where daylight saving time is not in effect:

- Honolulu, Hawaii, United States: UTC−10; 16:00 on Monday
- Toronto, Ontario, Canada: UTC−05; 21:00 on Monday

The time-zone adjustment for a specific location may vary because of daylight saving time. For example New Zealand, which is usually UTC+12, observes a one-hour daylight saving time adjustment during the Southern Hemisphere summer, resulting in a local time of UTC+13.

Time zone conversions

Conversion between time zones obeys the relationship

"time in zone A" − "UTC offset for zone A" = "time in zone B" − "UTC offset for zone B",

in which each side of the equation is equivalent to UTC. (The more familiar term "UTC offset" is used here rather than the term "zone designator" used by the standard.)

The conversion equation can be rearranged to

"time in zone B" = "time in zone A" − "UTC offset for zone A" + "UTC offset for zone B".

For example, what time is it in Los Angeles (UTC offset= −08) when the New York Stock Exchange opens at 09:30 (−05)?

time in Los Angeles = 09:30 − (−05:00) + (−08:00) = 06:30.

In Delhi (UTC offset= +5:30), the New York Stock Exchange opens at

time in Delhi = 09:30 − (−05:00) + (+5:30) = 20:00.

These calculations become more complicated near a daylight saving boundary (because the UTC offset for zone X is a function of the UTC time).

Nautical time zones

Since the 1920s a nautical standard time system has been in operation for ships on the high seas. Nautical time zones are an ideal form of the terrestrial time zone system. Under the system, a time change of one hour is required for each change of longitude by 15°. The 15° gore that is offset from GMT or UT1 (not UTC) by twelve hours is bisected by the nautical date line into two 7.5° gores that differ from GMT by ±12 hours. A nautical date line is implied but not explicitly drawn on time zone maps. It follows the 180th meridian except where it is interrupted by territorial waters adjacent to land, forming gaps: it is a pole-to-pole dashed line.[9] [10] [11]

A ship within the territorial waters of any nation would use that nation's standard time, but would revert to nautical standard time upon leaving its territorial waters. The captain is permitted to change the ship's clocks at a time of the captain's choice following the ship's entry into another time zone. The captain often chooses midnight. Ships going in shuttle traffic over a time zone border often keeps the same time zone all the time, to avoid confusion about work, meal and shop opening hours. Still the time table for port calls must follow the land time zone.

Skewing of zones

Ideal time zones, such as nautical time zones, are based on the mean solar time of a particular meridian located in the middle of that zone with boundaries located 7.5 degrees east and west of the meridian. In practice, zone boundaries are often drawn much farther to the west with often irregular boundaries, and some locations base their time on meridians located far to the east.

Difference between sun time and clock time during daylight saving time:

DST observed DST no longer observed DST never observed

For example, even though the Prime Meridian (0°) passes through Spain and France, they use the mean solar time of 15 degrees east (Central European Time) rather than 0 degrees (Greenwich Mean Time). France previously used GMT, but was switched to CET (Central European Time) during the German occupation of the country during World War II and did not switch back after the war.[12] Similarly, prior to World War II, the Netherlands observed "Amsterdam Time", which was twenty minutes ahead of Greenwich Mean Time. They were obliged to follow German time during the war, and kept it thereafter. In the mid 1970s the Netherlands, as with other European states, began observing daylight saving (summer) time.

There is a tendency to draw time zone boundaries far to the west of their meridians. Many of these locations also use daylight saving time. As a result, in the summer, solar noon in the Spanish town of Muxia occurs at 2:37 pm by the clock. This area of Spain never experiences sunset before 6:00 pm local time even in midwinter, despite its lying more than 40 degrees north of the equator. Near the summer solstice, Muxia has sunset times similar to those of Stockholm, which is in the same time zone and 16 degrees further north.

A more extreme example is Nome, Alaska, which is at 165°24′W longitude—just west of center of the idealized Samoa Time Zone (165°W). Nevertheless, Nome observes Alaska Time (135°W) with DST so it is slightly more than two hours ahead of the sun in winter and over three in summer.[13] Kotzebue, Alaska, also near the same meridian but north of the Arctic Circle, has an annual event on 9 August to celebrate *two* sunsets in the same 24-hour day, one shortly after midnight at the start of the day, and the other shortly before midnight at the end of the day.

Also, China extends as far west as 73°34′E, but all parts of it use UTC+08:00 (120°E), so solar "noon" can occur as late as 15:00 in western portions of China such as Xinjiang and Tibet.

Daylight saving time

Many countries, and sometimes just certain regions of countries, adopt daylight saving time (also known as "Summer Time") during part of the year. This typically involves advancing clocks by an hour near the start of spring and adjusting back in autumn ("spring" forward, "fall" back). Modern DST was first proposed in 1907 and was in widespread use in 1916 as a wartime measure aimed at conserving coal. Despite controversy, many countries have used it off and on since then; details vary by location and change occasionally. Most countries around the equator do not observe daylight saving time, since the seasonal difference in sunlight is minimal.

Additional information

- France has twelve time zones including those of Metropolitan France, French Guiana and numerous islands, inhabited and uninhabited. Russia has nine time zones (and used to have 11 time zones before March 2010), eight contiguous zones plus Kaliningrad exclave on the Baltic Sea. The United States has ten time zones (nine official plus that for Wake Island and its Antarctic stations; no official time zone is specified for uninhabited Howland Island and Baker Island). Australia has nine time zones (one unofficial and three official on the mainland plus four for its territories and one more for an Antarctic station not included in other time zones). The United Kingdom has eight time zones for itself and its overseas territories. Canada has six official time zones. The Danish Realm has five time zones.
- In terms of area, China is the largest country with only one time zone (UTC+08). China also has the widest spanning time zone. Before 1949, China was separated into five time zones.
- Stations in Antarctica generally keep the time of their supply bases, thus both the Amundsen-Scott South Pole Station (U.S.) and McMurdo Station (U.S.) use New Zealand time (UTC+12 southern winter, UTC+13 southern summer).
- The 27°N latitude passes back and forth across time zones in South Asia. Pakistan: +05:00, India +05:30, Nepal +05:45, +05:30, Myanmar +06:30. This switching was more odd in 2002, when Pakistan enabled daylight saving time. Thus from west to east, time zones were: +06:00, +05:30, +05:45, +05:30, +08:00, +06:00, +05:30 and +06:30.
- Because the earliest and latest time zones are 26 hours apart, any given calendar date exists at some point on the globe for 50 hours. For example, April 11 begins in time zone UTC+14 at 10:00 UTC April 10, and ends in time zone UTC−12 at 12:00 UTC April 12.
- There are 22 places where three or more time zones meet, for instance at the tri-country border of Finland, Norway and Russia. 28 countries present such triple points, with China being the most present (in 13 of the 22 triple points). Then come India (7), Russia, India and Afghanistan (4).
- There are 40 time zones. This is due to fractional hour offsets and zones with offsets larger than 12 hours near the International Date Line as well as one unofficial zone in Australia. See the list of time zones.
- The largest time gap along a political border is the 3.5 hour gap along the border of China (UTC+08) and Afghanistan (UTC+04:30).
- One of the most unusual time zones is the Australian Central Western Time zone (CWST), which is a small strip of Western Australia from the border of South Australia west to 125.5°E, just before Caiguna. It is 8¾ hours ahead of UTC (UTC+08:45) and covers an area of about 35,000 km^2, larger than Belgium, but has a population of about 200. Although unofficial, it is universally respected in the area—without it, the time gap in standard time at 129°E (the WA/SA border) would be 1.5 hours. See Time in Australia.

Computer systems and the Internet

Computer operating systems include the necessary support for working with all (or almost all) possible local times based on the various time zones. Internally, operating systems typically use UTC as their basic time-keeping standard, while providing services for converting local times to and from UTC, and also the ability to automatically change local time conversions at the start and end of daylight saving time in the various time zones. (See the article on daylight saving time for more details on this aspect).

Web servers presenting web pages primarily for an audience in a single time zone or a limited range of time zones typically show times as a local time, perhaps with UTC time in brackets. More internationally-oriented websites may show times in UTC only or using an arbitrary time zone. For example the international English-language version of CNN includes GMT and Hong Kong Time,[14] whilst the US version shows Eastern Time.[15] US Eastern Time and Pacific Time are also used fairly commonly on many US-based English-language websites with global readership. The format is typically based in the W3C Note "datetime".

Email systems and other messaging systems (IRC chat, etc.)[16] time-stamp messages using UTC, or else include the sender's time zone as part of the message, allowing the receiving program to display the message's date and time of sending in the recipient's local time.

Database records that include a time stamp typically use UTC, especially when the database is part of a system that spans multiple time zones. The use of local time for time-stamping records is not recommended for time zones that implement daylight saving time due to the fact that once a year there is a one hour period when local times are ambiguous.

Operating systems

Unix

Most Unix-like systems, including Linux and Mac OS X, keep system time as UTC (Coordinated Universal Time). Rather than having a single time zone set for the whole computer, timezone offsets can vary for different processes. Standard library routines are used to calculate the local time based on the current timezone, normally supplied to processes through the TZ environment variable. This allows users in multiple timezones to use the same computer, with their respective local times displayed correctly to each user. Time zone information most commonly comes from the IANA time zone database. In fact, many systems, including anything using the GNU C Library, can make use of this database.

Microsoft Windows

Windows-based computer systems prior to Windows 2000 used local time, but Windows 2000 and later can use UTC as the basic system time.[17] The system registry contains time zone information that includes the offset from UTC and rules that indicate the start and end dates for daylight saving in each zone. Interaction with the user normally uses local time, and application software is able to calculate the time in various zones. Terminal Servers allow remote computers to redirect their time zone settings to the Terminal Server so that users see the correct time for their time zone in their desktop/application sessions. Terminal Services uses the server base time on the Terminal Server and the client time zone information to calculate the time in the session.

Programming languages

Java

While most application software will use the underlying operating system for timezone information, the Java Platform, from version 1.3.1, has maintained its own timezone database. This database will need to be updated whenever timezone rules change. Sun provides an updater tool for this purpose.[18]

As an alternative to the timezone information bundled with the Java Platform, programmers may choose to use the Joda-Time library.[19] This library includes its own timezone data based on the IANA time zone database.[20]

JavaScript

There is very little in the way of timezone support for JavaScript. Essentially the programmer has to extract the UTC offset by instantiating a time object, getting a GMT time from it, and differencing the two. This does not provide a solution for daylight saving variations.

PHP

The DateTime objects and related functions have been compiled into the PHP core since 5.2. This includes the ability to get and set the default script timezone, and DateTime is aware of its own timezone internally. PHP.net provides extensive documentation on this.[21] As noted there, the most current timezone database can be implemented via the PECL timezonedb.

Python

The standard module datetime stores and operates on the timezone information class tzinfo. The third party pytz module provides access to the full IANA time zone database.[22] Negated time zone offset in seconds is stored time.timezone and time.altzone attributes.

Smalltalk

Each Smalltalk dialect comes with its own built-in classes for dates, times and timestamps, only a few of which implement the DateAndTime and Duration classes as specified by the ANSI Smalltalk Standard. VisualWorks provides a TimeZone class that supports up to two annually recurring offset transitions, which are assumed to apply to all years (same behavior as Windows time zones). Squeak provides a Timezone class that does not support any offset transitions. Dolphin Smalltalk does not support time zones at all.

For full support of the tz database (zoneinfo) in a Smalltalk application (including support for any number of annually recurring offset transitions, and support for different intra-year offset transition rules in different years) the third-party, open-source, ANSI-Smalltalk-compliant Chronos Date/Time Library is available for use with any of the following Smalltalk dialects: VisualWorks, Squeak, Gemstone, or Dolphin.[23]

Databases

Some databases allow storage of a datetime type having time zone information. The SQL standard defines two standard time data types:

TIMESTAMP WITH TIME ZONE

TIMESTAMP WITHOUT TIME ZONE

However, the standard has a somewhat naive understanding of time zones. It generally assumes a time zone can be specified by a simple offset from GMT. This causes problems when trying to do arithmetic on dates which span daylight saving time transitions or which span political changes in time zone rules.

Oracle

Oracle Database is configured with a database time zone, and connecting clients are configured with session time zones. Oracle Database uses two data types to store time zone information:

TIMESTAMP WITH TIME ZONE

Stores date and time information with the offset from UTC

TIMESTAMP WITH LOCAL TIME ZONE

Stores date and time information with respect to the dbtimezone (which cannot be changed so long as there is a column in the db of this type), automatically adjusting the date and time from the stored time zone to the client's session time zone.

PostgreSQL

PostgreSQL uses the standard SQL data types but tries to impose an interpretation which avoids the problems described above.

TIMESTAMP WITH TIME ZONE

Stores date and time in UTC and converts to the client's local time zone (which could be different for each client) for display purposes and conversion to other types.

TIMESTAMP WITHOUT TIME ZONE

Stores date and time without any conversion on input or output. When converting to a TIMESTAMP WITH TIME ZONE, interprets it according to the client's local time zone.

TIME WITH TIME ZONE

Stores time of day together with a UTC offset in which it is to be interpreted.

TIME WITHOUT TIME ZONE

Stores time of day without any time zone specification.

Time zones in outer space

Orbiting spacecraft typically experience many sunrises and sunsets in a 24-hour period, or in the case of Apollo program astronauts travelling to the moon, none. Thus it is not possible to calibrate time zones with respect to the sun, and still respect a 24-hour sleep/wake cycle. A common practice for space exploration is to use the Earth-based time zone of the launch site or mission control. This keeps the sleeping cycles of the crew and controllers in sync. The International Space Station normally uses Coordinated Universal Time (UTC).

Timekeeping on Mars can be more complex, since the planet has a solar day of approximately 24 hours and 39 minutes, known as a sol. Earth controllers for some Mars missions have synchronized their sleep/wake cycles with the Martian day, because solar-powered rover activity on the surface was tied to periods of light and dark. The difference in day length caused the sleep/wake cycles to slowly drift with respect to the day/night cycles on Earth, repeating approximately once every 36 days.

See also

- Daylight saving time
- ISO 8601
- Jet lag
- List of time zone abbreviations
- List of time zones by country
- Lists of time zones
- Metric time
- Sunpendulum
- Time by country
- World clock

The control panel of the Time Zone Clock in front of Coventry Transport Museum.

References

[1] Latitude and Longitude of World Cities http://www.infoplease.com/ipa/A0001769.html

[3] Economics of Time Zones (http://bpp.wharton.upenn.edu/mawhite/Papers/TimeZones.pdf) PDF (1.89 MB)

[6] Quirico Filopanti (http://www.scienzagiovane.unibo.it/english/scientists/filopanti-1.html) from scienzagiovane, Bologna University, Italy.

[7] Gianluigi Parmeggiani (Osservatorio Astronomico di Bologna), The origin of time zones (http://www.iav.it/planetario/didastro/didastro/english.htm)

[8] The Astronomical Almanac 1983, US Government Printing Office (Washington) and Her Majesty's Stationery Office (London), page B4.

[9] Bowditch, Nathaniel. *American Practical Navigator*. Washington: Government Printing Office, 1925, 1939, 1975.

[10] Hill, John C., Thomas F. Utegaard, Gerard Riordan. *Dutton's Navigation and Piloting*. Annapolis: United States Naval Institute, 1958.

[11] Howse, Derek. *Greenwich Time and the Discovery of the Longitude*. Oxford: Oxford University Press, 1980. ISBN 0-19-215948-8.

[17] GetSystemTime function (Windows) (http://msdn.microsoft.com/en-us/library/windows/desktop/ms724390(v=vs.85).aspx)

[23] Chronos Date/Time Library (http://www.chronos-st.org/)

External links

- LEGAL TIME 2013 (http://www.itu.int/dms_pub/itu-t/opb/sp/T-SP-LT.1-2013-OAS-PDF-E.pdf)
- W3C Note Datetime (http://www.w3.org/TR/NOTE-datetime)
- US Official Time Clock (http://www.time.gov/timezone.cgi?UTC/s/0/java) Java-enabled clock to graphically display night and day around the globe.

Article Sources and Contributors

Sewickley_Heights,_Pennsylvania *Source*: http://en.wikipedia.org/w/index.php?title=Sewickley_Heights%2C_Pennsylvania *Contributors*: Acntx, Canadian2006, David H. Flint, Dpetley, Gerry D, Hmains, Jackonferry, Jllm06, Jmlk17, Joseph Solis in Australia, Ken Gallager, Lightmouse, Magioladitis, Marketdiamond, Plastikspork, Ram-Man, VerruckteDan, 10 anonymous edits

Allegheny_County,_Pennsylvania *Source*: http://en.wikipedia.org/w/index.php?title=Allegheny_County%2C_Pennsylvania *Contributors*: AXRL, Acntx, AdamRetchless, Adavidb, Ahoerstemeier, Ale jrb, Alekjds, AlistairMcMillan, All Is One, Anderfreude, Andycjp, Anonymous56789, AshyLarry, Atif.t2, BRG, Backspace, Badbilltucker, Bamyers99, Bender235, Berean Hunter, Bkonrad, Blargh29, Bmstephany, Bobblewik, Bobby H. Heffley, Boothy443, Brybry1999, Buddha24, CanisRufus, Caponer, Catwheel76, Cavalcabo, Ccacsmss, Charleson, Choess, Chooch51, Choywohm, Citterio, ClairSamoht, Clarkbhm, Closedmouth, Coemgenus, CommonsDelinker, Connormah, Cornellrockey, Crazypaco, D6, DanaSheehan, Db099221, Deedee19482, Desertsky85451, Djkimmons, Dneubert, Download, DrRon, Dthomsen8, Dwscomet, EATC, Edward, Everyking, Fat&Happy, Frankbaseball27, Franklincrew, Fredrik, Gaius Cornelius, GamblinMan22, Gehirn, Gerry D, Gimboid13, Glöckner der Große, Good Olfactory, Great Scott, Heff01, Hephaestos, Hmains, Hu12, Hyperfast, Infamousjre, IsWayneBradygonnahavetosmackabitch, JeffersonHills, Jim.henderson, Jmlk17, Jwh5116, KAB-J, Ken Gallager, Kevin Myers, Lady Mondegreen, Lakeererules1, Langhorner, Lavallen, Leslie Mateus, Lightmouse, Lotje, Lprmesia, Magioladitis, Maile66, Major Danby, Malepheasant, Mareino, Marketdiamond, Milton Stanley, Monegasque, Muriel Gottrop, NameThatWorks, Nevermore27, Niagara, Nick Number, Nk, Notyourbroom, Numbo3, Nyttend, Ohnoitsjamie, Omnedon, PBRTV, PSUMark2006, Paris1127, Paschmitts, Peetlesnumber1, Pfly, PghChampionsCity, Piotrus, Pip2andahalf, Polmandc, Postdlf, Pramshaw, Pwu2005, Ram-Man, Rapsar, Rayoflight278, ReadQT, Rich Farmbrough, RichardPgh, Rmmst22, Rossdegenstein, Ruhrfisch, SPUI, Safiel, Schzmo, Scillystuff, Search4Lancer, Skeetidot, Skinsmoke, Smallbones, Solar-Wind, Stansult, Steelcityartst, Swpa911, TUF-KAT, TastyPoutine, Template namespace initialisation script, Tetulun, Thunderbunny, Timothy Thomasen, TooPotato, Vanished user 194difuh2ruhqwdoinxojakdjncno234r, VerruckteDan, VitaleBaby, Wars, Whhalbert, Whpq, WillC, WilliamJE, Wmahan, Worldenc, Yoshi Canopus, Zoe, 179 anonymous edits

Pennsylvania *Source*: http://en.wikipedia.org/w/index.php?title=Pennsylvania *Contributors*: "Country" Bushrod Washington, -Midorihana-, 00dick, 09kmeyer, 15turnsm, 16@r, 1Effect, 1exec1, 2001:558:E0:305:0:5EFE:AA3:1DC2, 2602:306:BC74:6EE0:5D85:2E28:6F77:C2D8, 2D, 6thgradequeenlegends, 7&6=thirteen, A. Parrot, A104375, A2Kafir, A8UDI, ACW1984, ARUenergy, AaronW, Abigail-II, Academic Challenger, Acalamari, Achmelvic, Achowat, Acroterion, AdRock, Adam Rock, AdamRetchless, Addihockey10, AdjustShift, AdultSwim, Aef711, Aeusoes1, Agent L, Aglebagel, Agn106, Agricolae, Ahassan05, Ahoerstemeier, Aint, Airwolf, Aivazovsky, Ajraddatz, Akhristov, Al tally, Alakazam, AlanBarrett, Alanscottwalker, Alansohn, Ale jrb, AlexiusHoratius, Alexwcovington, Ali, All Is One, AllStarSun1, Allstarecho, Alpha 4615, Alphaboi867, Alteripse, Amazonien, Analtrollgoatse, Anb racsh, Andonic, AndreaPersephone, Andrew Rodland, Andrew845, Andrewcmcardle, Andrewman327, Andris, Andy Marchbanks, Andy120290, Angelo.romano, Angr, Angrysockhop, Angusmclellan, Animum, Anniegirl207, Antandrus, Anthony71, Antodav2007, Antonio Lopez, Anuvar, Appraiser, Arakunem, AshishG, Aslayerfan, Astarte, Asturianu, AtheWeatherman, Atrix20, AuburnPilot, Aude, Aure1ius9, AussieLegend, Aventinus, Avoided, AwamerT, Axtreampeep, Ayudante, B64, BAPACop, BD2412, BSveen, Bab123, BabuBhatt, Bacteria, Balcer, Bananas4life123, Bantman, Barek, Barneca, Barryblack, Baseballnum5, Bbert1994, Bbert94, Beetstra, Beezhive, Behanna, Beland, Belligero, Beloki, BenBaker, Bender235, Benesch, Benhealy, Benjh40, Bentley4, Berean Hunter, Berger+Jelly, Berniethomas68, Besidesamiracle, Betta18, Bhess, Big iron, BigBoyRubio, Bigdave55, Bigfootjosh, Bility, BillFlis, Billinghurst, Binary TSO, Bitoffish, BjKa, Bjrooney, Bkell, Bkonrad, BlackRaspberry, Blameless, Blanchardb, Bletch, BloodRevenge45, Blumster, BlurTento, Bman1530, Bmv 1978, Bneepgrim, Bob5171996, Bobak, Bobblewik, Bobm121, Bobo the Talking Clown, Bobo192, Bobrayner, Bocianski, Bomac, Bongwarrior, Boothy443, BoringHistoryGuy, Bpmullins, Brian1979, Bridge Partner, Brighterorange, Brion VIBBER, Broken Claw, Bronayur, BrownHairedGirl, BruceDLimber, Bruno098, Bryan Derksen, Brybry1999, Buaidh, Bubba hotep, Buddha24, Buf7579, Buggie111, Burnout (video game), Burnwelk, Bushcarrot, Buster2058, Bwilkins, C.Fred, C6541, C777, CA387, CO, CORNELIUSSEON, CPacker, CWii, Cabe6403, Cadiomals, Caesar Rodney, Calabe1992, Calibaby, Callanecc, Caltas, Cameron8782, Campdavid, Can't sleep, clown will eat me, CanadianLinuxUser, CanisRufus, Cannedy22, Caponer, Capricorn42, Carfreak2128, Carlos321, Carnildo, Castncoot, Catdude, Catgut, Cburnett, Ccacsmss, Cchow2, Ccscott, Cessna38671, Ceyockey, Cflm001, Chanheigeorge, CharlieGrammar, CharlotteWebb, Chartran, Cherkash, Chris the speller, ChrisGriswold, ChrisGualtieri, ChrisRuvolo, Chrislk02, Christy747, Cisum6cbb, Civil Engineer III, Cizzle107, Ck lostsword, Ckatz, ClairSamoht, Clarkbhm, Clepper26, Clockworkclock, Closedmouth, Cmcfarland, Coemgenus, Colonies Chris, Cometstyles, CommonsDelinker, Coneslayer, Confiteordeo, Conk 9, Conradsbrother, Conversion script, Cool Blue, Corpx, CottrellS, Covington, Crazypaco, Cremepuff222, Cresix, Cribbswh, Crunchy Numbers, Crunchyman, Crystal whacker, Cs8782, Csigabi, Csoldner187, Cst17, Ctrabs74, Cultofpj, Cwp10, CylonCAG, D. Recorder, D.whitey19, D6, DARTH SIDIOUS 2, DBaba, DBishop1984, DCEdwards1966, DIDouglass, DJ Clayworth, DJ WikiBob, DK4, DLinth, DVD R W, Dabbler, Dabomb87, Dacoslett, Daduzi, DagnyB, Dale Arnett, Damicatz, Dancer1000, DanielCD, Danielvalle2222, Danny, Darkwind, Darth Panda, Darthgriz98, Darthveda, Dasani, Dave6, David Justin, Davideus86, Dayfled2, Dbulwink, Dddstone, DeanHinnen, Debresser, December12AC, Deflective, DeltaQuad, DemocraticLuntz, DerHexer, Dfitzgerald, Dhp1080, Dhwani1989, Diannaa, DigitalNinja, Dirtykike3728, Discogeek, Discospinster, Djinn112, Djorion, Dkdantastic, Dlae, Dleav, Dneubert, Doc glasgow, DocWatson42, Docgaeta, Docu, Doggy smell, Dopdat313, Dormant25, Dormantfascist, DoubleBlue, Dough4872, Doulos Christos, Download, Dp462090, Dp67, Dppowell, Dr. Blofeld, Draggleduck, Dravecky, Drbreznjev, Drilnoth, Drmies, Drumguy8800, Dspserpico, Dthomsen8, Dudtz, DuncanH, Durakote, Duty gurl, Dwo, Dwscomet, Dycedarg, Dysepsion, Dysmorodrepanis, Dysprosia, E-trail, EATC, EXe I Reaper, Eagle4000, EaglesFanInTampa, Earl0060, East of Borschov, Eastlaw, Ecjmartin, Eclecticology, Ed g2s, Eddiezeppelin, Eddster, Editer001, EditorPIN, Edivorce, Eeekster, EikwaR, Ekren, Elcaptain87, Elcaset, Eldonjon72, Elipongo, Elockid, Emachinest3256, Emc2, Emhoo, Enam Esru, EncycloPetey, Enviroboy, Epbr123, Epicadam, Epolk, Epsteada, ErgoSum88, Eric-Wester, EricBarker, Ertemplin, Esanchez7587, Escape Orbit, Eu.stefan, EurekaLott, Euryalus, Everyguy, Everyking, Evice, ExRat, Excirial, Extra999, FF2010, Fabant, Falcon8765, Fanghong, Farquaadhnchmn, Favonian, FecalSpam, Feezo, Female bodybuilder enthusiast, Fetchcomms, Feyer, Fg2, Fieldday-sunday, Filll, Finavon, Finn-Zoltan, Firdingerdingerdus, Firsfron, FisherQueen, Fishing, Flami72, Flatterworld, Flonto, Fluffybunny6969, Fluri, Flyguy33, Flyinglemurs12345, Fnlayson, Focus, Fourthords, Fr. Rizzo, Fran Rogers, FrancoGG, Frankenpuppy, Fratrep, Freddiem, Fremsley, Frietjes, Fruggo, FuelWagon, Fufthmin, Fuhghettaboutit, Funnyhat, Furado, Fuzheado, Fuzzy510, Fyunck(click), Fyver528, Fyyer, Fæ, GDonato, Gaius Cornelius, GamblinMan22, Gatekeeper 21, Gatkinso, Gauss, Gays are awfully disgusting, Gdr, GenuineMongol, Geologyguy, Georgian, Geriguiaguiatugo, Gerry D, Gertlex, Giants27, Gif32, Gigidygigidygoo, Gilliam, Gimmetrow, GinaKin, Glane23, Gogo Dodo, GoldRingChip, Goldfishbutt, Goobergunch, Good Olfactory, Goodnightmush, Gota1respect1, GraemeMcRae, Grafen, Graham87, GrapedApe, Great Scott, GreatLakesdemocracy, GreenGourd, Greengamer4life, GregAsche, GregU, Grim23, Ground Zero, Grsz11, Gstamets, Guardian of the Rings, Guitarman960, Gurch, Gustavb, Gwernol, H2eddsf3, HHRVB, HalfShadow, Hammertime, HangingCurve, Hanky25, Happygillmore192, Happyplace123456, Harjk, Harmil, Harpiegirl6, Hasek is the best, Hay413, Haymaker, Hdt83, Heaz07, Hec7, Heegoop, Hekerui, Helllord33, Hellmaiden, Hellomynameis.....blah, Hemanshu, Herald-D-Junior, HexaChord, Hhand, Highvale, Hillsboro, Historian932, History21, HkCaGu, Hmains, Hmrox, Hobophobe, HokieRNB, Holyreaper15, Horologium, Horselover90909, Hotrodjim, Houshyar, Howrealisreal, Hto9950, Hu12, Hubstack, Hucz, Huduuthink, HunterRex, Hurriquake, Hushpuckena, HuskyHuskie, Hut 8.5, Hvn0413, Hydrogen Iodide, I see the truth inside your eyes Yo, IBook of the Revolution, IW.HG, Icangityonumba, Ictlogist, Igiffin, Igrek, Ikescs, Immunize, InMooseWeTrust, Indosauros, InfoTweak, Iridescent, Irishguy, IronGargoyle, Ishkur88, Isis, Isopropyl, Ivirivi00, Ixfd64, J.delanoy, JBKramer, JForget, JFreeman, JHunterJ, JNW, JYi, JaMikePA, Jackollie, JackyR, Jacob.jose, Jadeddissonance, Jag164, Jajhill, Jaksmata, James086, JamesAM, Jan eissfeldt, Janiediane, JanusK, Jared Kofsky, Jasonsk287, Jawny80, Jc2sexay, Jdheim, Jdjd021, Jebus989, JediScougale, Jeepday, Jeff G., Jeffq, Jengod, Jennavecia, Jensen627, Jerwal00, JesseRafe, JetLover, Jgera5, Jgirata, Jgoelz102, Jhendin, Jiang, JimIrwin, JimVC3, Jimi magic, Jimk1943, JinJian, Jkarns, Jklin, Jll, Jmboothe, Jmpenzone, Jncraton, JoanneB, JoeSmack, Joefromrandb, Johanthesaint, John Cline, John K, JohnInDC, Jojhutton, Jon Harald Søby, Jonas Salk, Jonesey, Jordzlaaaak, Joseph Solis in Australia, Jossi, Joyous!, Jprg1966, Jpz109, Jschnalzer, Jtwnirish, Juicycat, Juliancolton, Jultemplet, Jusdafax, Jxl180, Karam.Anthony.K, Karenjc, Katalaveno, Kbh3rd, Keahi212, KeithB, KeithH, Keizersnl, Ken Gallager, Kerotan, Kether83, Kggantz, Khatru2, Kihiu, Kildruf, Killervogel5, King of Hearts, Kingbling11, Kingxxx, Kinomakoto, Kintetsubuffalo, KirbyMaster14, Kjtermaat, Klemen Kocjancic, KnightLago, KnoxSGT, Kowalmistrz, Kozuch, Kpk8, Kralizec!, Kramden4700, Krazychris81, Krikkit261, Krugerhlu, Ks0stm, Kscrt584, Ksy92003, Kungfuadam, Kurt Jansson, Kurtbw, Kuru, Kuyabribri, Kwamikagami, L Kensington, L1A1 FAL, LA2, LAX, LDV, LabFox, Laikalynx, Lalalajerk, Lalalosurr, LarryJeff, Lasunncty, Latitude0116, Latka, Laureapuella, Lawilkin, Legoktm, Lethargy, Leuko, Levi42, Levineps, Liface, Lightmouse, LikeLakers2, LilHelpa, Lilac Soul, Lilpinoy 82, Lindsayandrews7, Linelor, LinkinPark9297, Little Mountain 5, LittleOldMe, LizardJr8, Locquincer, Logophile, LoonyLeif, Looper5920, Loopism, Lou Sander, LouI, Lthompson1, Ltwin, LuK3, Luckypenutnose, LuftWaffle0, LuigiManiac, Lukasz Lukomski, Luna Santin, Luuva, Luvit200, M C Y 1008, MJCdetroit, MJDTed, MK8, MMonk Cowpens, MPF, MPerel, Mac, Macrowiz, Maelnuneb, Magicusa, Magister Mathematicae, Magog the Ogre, Majorly, Malcolm Farmer, Malleus Fatuorum, Malo, Man17018, Mandarax, Manuel Trujillo Berges, Mapetite526, Marcsin, Mareino, Marek69, Marianocecowski, Markco1, Marnanel, MaroonGray213, Martarius, Martin451, Martinp23, Martinribadeneira, Master of Puppets, Mathpianist93, Matthardingu, Mattww123456789, Mausy5043, Mav, Maximus Rex, Maxtremus, Mayumashu, Mbertsch, McCart42, Mean as custard, Meekywiki, Megan1967, MeltBanana, Meredyth, Merlinsorca, Metropolitan90, Michael J, MightyWarrior, Mikibacsi1124, Millero10, MiloKral, Mimicatty, Minesweeper, Minniegril8253, Miranda, Mister Pinball, Mks12, Mlaffs, Mls393, Mm40, Mmiklas, Mo0, Monnicat, Mooby the Golden Sock, Moralis, Mordgier, Morwen, Mosmof, MountainRail, Mouse Nightshirt, Moverton, Moxy, Mr Accountable, Mr45acp, MrPMonday, Mrees1997, Mrmcdonnell, Mrreallycool, Mrsmith93309, Mrwojo, Ms2ger, Mulder416, MusicianChick, Mvincec, Myanw, Myasuda, Myownworst, NHRHS2010, NJGW, Nacho Librarian, Nahallac Silverwinds, Nairobiny, Nakon, Narsil27, Nat682, NawlinWiki, Neelvk, Neko-chan, Neodem, Neonblak, Neutrality, Nev1, NewEnglandYankee, Newport Beach, Newsaholic, Niagara, Nicholasnice, Nick, Nickbinman, Nightscream, Nihiltres, Nikinikolananov, Nikopoley, NinjaCharlie, Ninly, Nishkarshs, NisiFara, NitRav, Nivix, Noctibus, Non-Dairy Creamer, Nostory, Npeters22, Nphs2010, Nrafter530, Nsaa, Nubiatech, NuclearWarfare, Nv8200p, Nyttend, OHWiki, Ohconfucius, Ohnoitsjamie, Okiefromokla (old), OldsVistaCruiser, Olessi, Olivier, OllieFury, Omnionline, Oneboy, Onepebble, Onorem, Oo64eva, Oradej, Orphic, Osarius, Osio567, Ouishoebean, OwenBlacker, OwenX, OxyMoronMinusOxy, Oxymoron83, P. S. F. Freitas, P4k, PA Vacation, PAWiki, PCHS-NJROTC, PKM, Padvi, Pandaslayer69, Papercrab, Parkwells, Parsecboy, Pascal, Pastricide, PatGallacher, Patar knight, Patrickneil, Paulmcdonald, Pdakeyboardplaya, Pearle, Peetlesnumber1, Pekinpekin, Penguinluvr22888, Penndot1, PennsylvaniaGov, PeterSymonds, Peteritism, Pgan002, Pgk, Phantom784, Phantomsteve, Pharaoh of the Wizards, Phgao, PhiladelphiaCVB, Philip Trueman, Philippe, Phillies1fan777, Phillip J, PhillyPartTwo, Piano non troppo, Picnicbystream, Pigman, Pikeweatherman, Piledhigheranddeeper, Pilotguy, Pilover819, Pinethicket, Pitt99, Piuslxix, Planckton, Pne, Pnkrockr, Poccil, Poeloq, Polaron, Polarscribe, Pollinator, Postdlf, Praline97, Prestonmag, Ps0313, Psy guy, Purpleblack123, Pwh4130, Pwt898, Pwu2005, Pzavon, Qatter, Qqqqqq, Quantpole, Question12, Quilokos, Qwertyus, Qwyrxian, Qxz, R'n'B, R3D3Y3S27, RBBrittain, RJN, RJaguar3, RMHED, RNLion, RUL3R, Radon210, RaggedBiscuit, Rah126, Ram-Man, RamBow, Ranieri66, Ravedave, Ray Glock-Grueneich, RayBirks, Rayoflight278, Razorflame, Reagank, Real decimic, Realm of Shadows, Recognizance, Red Winged Duck, RedHillian, RedShiftPA, RedWolf, Redneb, Reedy, Retired username, Rettetast, Revas, Reywas92, Rfc1394, Rfl, Rgrof, Rhashan1, Rhatsa26X, Rich Farmbrough, RichardF, Rick Block, Ricky81682, Rickyburger15, RingtailedFox, Rjd0060, Rjwilmsi, Rmhermen, Rmosler2100, RobertG, Robhmac, Romanm, Ronark, Ronhjones, Ronnocian159753, Roodog2k, Root Beers, Rorygibbensrolph, Rossdegenstein, Rpfitzgerald, Rreagan007, Rrius, Rsherman40, Rsieg, Rufus843, Ruhrfisch, Ruok9c, Rwboa22, Ryan wallack, Ryulong, S-1-5-7, SB Johnny, SNIyer12,

STEPHENFINAFROCK, Sad gurl, Salvioshia, Sam Staton, Sammy Houston, Samuel Blanning, Sanbeg, Sango123, Sannse, Santa-411, Saopaulo1, SatyrTN, Saurarity, Savatar, Saz127, Sbressi, Scanlan, Scientizzle, Scjessey, Scope creep, Scorpio1981, Scranton215, Scsmatt, Sean.hoyland, Search4Lancer, Seb az86556, SebastianHelm, Sfan00 IMG, Sfmontyo, Sgoldman10, Shanes, Shark96z, Shaunv123, Shiglhil, Shiquasa, Shoaler, Shouta, Shrommer, Shuffdog, Sietse Snel, Sillyserhan, Silvermoonx, SimonP, Skabat169, Skarebo, Skaterfunk, SkerHawx, Skiasaurus, Skizzik, Skunkboy74, Sky Attacker, Slambo, Slipperyweasel, Sloman, Sloshmo, Slushy9, Slvrstn, Smarkflea, Smith03, Snellmann, Snow Blizzard, Snoyes, SoCalSuperEagle, Socially Isolated Hamster, Solipsist, Some jerk on the Internet, SomeFreakOnTheInternet, SomeUsr, Sonarclawz, Sonicobbsessed, Sourpatchkids666, South Bay, South Philly, Sp33dyphil, Sparkhurst17, Spence737, Spiesr, Spikebrennan, Spitfire, SpookyMulder, Spotteddogsdotorg, Sprigdog, Spyder Monkey, Spydrlink, Spyneyes, Sriharsh1234, Sroulik, Sshikora, StAnselm, Standejb, StaticGull, Stbalbach, Steel, SteelMariner, Steelcityartst, SteffenB, Steinsky, Steve03Mills, SteveSims, Strannik, Strongbad1982, Suffusion of Yellow, Sullivan9211, Sunborn, SuperHamster, Superbeecat, Svgalbertian, Swanseamom, Swarm, SweetNeo85, Swollib, Sysm0921, T1980, TAnthony, TPIRFanSteve, TUF-KAT, Tangotango, TastyPoutine, Tbhotch, Tcncv, Tdmthomas, TecmoBo, Tedernst, Tedickey, Telekenesis, Template namespace initialisation script, Tesscass, Tetraedycal, Tgeairn, The Epopt, The Evil Spartan, The High Fin Sperm Whale, The Man in Question, The Thing That Should Not Be, The Transhumanist, The Utahraptor, The Wednesday Island, TheAlphaWolf, TheDJ, TheHoosierState89, TheRanger, Thebob000, Thegish, Thegreatdr, Thekrazykerzman, Thenameswill, Thesouthernhistorian45, Thetrollerstroll, Thingg, ThisguyYEA, Thisisbossi, Thomas Paine1776, Thunderbunny, Tide rolls, Tidying Up, Tiller54, TimBentley, Timcough, Timmyxcore666, Timneu22, Tinkey2, Tiptoety, Titoxd, Tiyang, Tlesher, Tlonca, Tnxman307, Tobby72, Tohd8BohaithuGh1, Tom Morris, TomStar81, Tomdobb, Tommy2010, TommyBoy, Tompie913, Tomtom, Toniokiv, Tony1, Toobevr1244, Toohool, TortoiseWrath, Towsonu2003, Tpbradbury, Tracer9999, Trafford09, Traforad, Tranquility2007, Travellingstone, Tresiden, Trimalchio, Triscuitface, Trugster, Trusilver, Truthanado, Tstrobaugh, Tunafizzle, Tuneguru, TwinCityIL, TwinsMetsFan, Twobestfriendz, Tyrol5, Ufwuct, Ukexpat, Ulmanor, Ulric1313, Ultimadesigns, Unity-Girl, Unyoyega, UsefulWikipedia, User5437, Username is shelby, Ussav, Uzusan, VB101, Vafthrudnir, ValleyRoad19046, Vanished user 194difuh2ruhqwdoinxojakdjncno234r, Vanished user ewfisn2348tui2f8n2fio2utjfeoi210r39jf, Vanished188, Vasiliymeshko, Vegaswikian, Veinor, Velvetron, Verrai, Versus22, Violinobaixo, Viridian, Volphy, Vpuliva, Vrenator, Vsmith, Vzbs34, WLRoss, WadeSimMiser, Wafulz, Wapcaplet, Wars, WashingtonWillie, Wavelength, Wayne Slam, Weatherman1126, Welsh, WereSpielChequers, Wereon, West Brom 4ever, Wester, Whatcanbrowndo, WhisperToMe, WikHead, Wiki Mateo, Wiki alf, WikiZorro, Wikibofh, Wikichanger22, Wikigirl98, Wikiwiji, Wild Wolf, Williamr10, Willking1979, Willydick, Wilsongirl00, Wimt, Wissgrl95, Witchzilla, Wknight94, Wmahan, Woohookitty, Woudloper, Wrightchr, Write On 1983, Wtmitchell, Wyattrahl, X292FirefighterX, Xb-70, Xfatmoe, Xyzzyva, YUL89YYZ, Yamamoto Ichiro, YankeeDoodle14, Yanksox, Yanwen, Yappy2bhere, Yarvin, Yayme23, YellowMonkey, Yilloslime, Yo its yamom, YodaRULZ, Yopure, Youngamerican, Your Mom3740, Yuenglingmaster, Zarcadia, Zavtrakat, Zeno22, Zereshk, Zimbabweed, Zoe, Zoicon5, Zokashi, Zooando, Zulitz, Zundark, Zzedar, Zzyzx11, Érico Júnior Wouters, პაატა შ., , 3001 anonymous edits

Population_density *Source*: http://en.wikipedia.org/w/index.php?title=Population_density *Contributors*: 1234r00t, 1297, 21655, 2mcm, 45Factoid44, 4halsop, A-giau, A. Parrot, A.Ou, Abhijitsathe, Achim Jäger, Addihockey10, Agentbla, Ahoerstemeier, Alansohn, AlexD, Alexjohnc3, Andre Engels, Andycjp, Anthonyz747, Aphaia, Apokryltaros, Applecharm, Aranel, Areology, Astuishin, Aude, Avicennasis, Aziz1005, B.d.mills, Barend, Barneca, Biederman, Blacklite, Blaircorrigan, Blanchardb, Bobblewik, Bogdangiusca, Br77rino, BrainyBabe, Brianlo747, Bricklayer, Brisvegas, BrokenSegue, Bsadowski1, Buchanan-Hermit, Bwilkins, CUSENZA Mario, Calabe1992, Candorwien, Catgut, Cdc, Cenarium, Chimpex, Ckatz, Cliff smith, Clovis Sangrail, Coffee, CommonsDelinker, Corpx, Courcelles, Cretog8, Cwgordon7, D. Cerv, DCEdwards1966, DL81292, DancingPenguin, Darth Panda, Davidcannon, Dbfirs, Dekisugi, Desiphral, Diaboli, Docu, Donarreiskoffer, DoorFrame, Dragoon17, Drrngrvy, Dusti, Dysprosia, E Pluribus Anthony, ERobson, Eagle4000, EdGl, Edgar181, Edward, Eeyore tim, Egg Centric, Eggman64, Eleassar777, Elekhh, Elmsat, Elockid, Eopsid, Epbr123, Eric Sandholm, EugeneZelenko, Excirial, Extra999, Eyedubya, Eyeofhorace, Farosdaughter, Figureskatingfan, Filemon, Francium12, Frozen4322, Func, GFW82, Garion96, Gcorral, GeeJo, Geremia, Gergis, Ghaly, Gidonb, Ginsengbomb, Glen, Gnosis1185, Gog, Gparalikidis, Graculus, Grafen, GrayFullbuster, Gryffindor, Gwabell, HGB, HalfShadow, Hallows AG, Hamedog, Helix pomatia, HenryLi, Hermitage17, Heron, Hitman012, Hoplophile, HornetMike, Hsm rocks ya right, Hypertall, Hégésippe Cormier, INeedANameThatIsFunny, Ibbn, Introvert, Iridescent, Island, J.delanoy, JNW, JUSTM3-X, Jacob.jose, Jake Spooky, Jakeburberry, JavierMC, Jaw959, Jdeamerr, Jds10912, Jeff G., Jim457, Joblio, JohnDoe0007, Johnuniq, Jonkerz, Jorgeposada14, Joseph Philipsson, Joseph Solis in Australia, Josh howson, Josisb, Jpk, Jprksh, Jrein59, Kaare, Karl-Henner, Karok, Keenan Pepper, Keilana, KenFehling, Ketiltrout, Kevin Gorman, Klparrot, KnowledgeOfSelf, Kr4ft, Kransky, Kurieeto, Kyle1278, L Kensington, Lakers, Latebird, LcawteHuggle, LeaveSleaves, Leks81, Leszek Jańczuk, Levineps, Liftarn, Lightmouse, Logose, Lotu, Lucas175, Lugia2453, LukeStewart, Luxborealis, Lytsio, MER-C, MJCdetroit, MaNeMeBasat, Mack2, Malcolmxl5, MapsMan, Marc Lacoste, Marek69, Mark Arsten, Mark J, Martial75, Materialscientist, Matthewcgirling, Maximus Rex, McSly, Melcombe, Mgnbar, MidgleyDJ, Mightymights, Miguel.mateo, Mindmatrix, Mintleaf, Missmarple, Moreschi, Motoxrampage101, Mrbuddhafreak, Mschlindwein, Mxn, NLikhov, NTK, Nagy, Nakon, Naval Scene, Neko-chan, Nentrex, Nepenthes, NewEnglandYankee, Newbyguesses, Nick C, Noah Salzman, Nunquam Dormio, Nyttend, Olly150, Olsonnnnn, Ongar the World-Weary, Oreo Priest, Orestimba49, PBP, Panicburststs, Patrick, Pax:Vobiscum, Pearle, Pengo, Pgan002, Phaust, PhilKnight, Philip Trueman, Physchim62, Piano non troppo, PierreAbbat, Pigman, Pinethicket, Piotrus, Pleasetouchme, Popsracer, Professionalsmartpeople1, Pumpmeup, Quasipalm, R Lowry, R'n'B, RLeopard, Radishes, Readyhairman, Red Slash, Remuel, Rev. John, ULC, RexNL, Rgdboer, Riana, Richard Warrington, Rmhermen, Romanm, Ronhjones, RoyBoy, Rushbugled13, Ryaelt, Ryan Valdés, SDC, SFK2, Samrager, SarekOfVulcan, Sarenne, Scohoust, Shizhao, Simmania, Simonkeuler.1245, Sin Harvest, Skinnyweed, SmthManly, Snowolf, Snowolfd4, Sonicyouth86, Spencer, Srasku, StaticGull, Ste, Steven Zhang, Steveprutz, Stormie, Student7, Sumahoy, Sunoco, Superm401, Surfbear335, Symode09, TGalt, TUKAKKAR, Tabitcorne2, TaintedMustard, TakuyaMurata, Tangotango, Tanzy97, Taykp, Tbhotch, Tellyaddict, Terence, The Angel of Islington, The Anome, The Determinator, The Hybrid, The Thing That Should Not Be, Thecaeser, Thehelpfulone, Thekohser, Theopolisme, Thermalite, Thucydides of Thrace, Thunder8, Tide rolls, Timeshifter, Tktktk, Tlc61, Tobias Galtieri, Tom harrison, Tomatoman180, Topbanana, Torchwoodwho, Tradewater, Tree Kittens, Trengarasu, Triesault, Tsxado, Tuanglen, Una Smith, Uncle Dick, Uuuussssseeeerrrrnnnnaaaammmmeeee, Vegard20, Violetriga, Vollex, Vranak, Waggers, Waldir, Wangi, Wasbeer, Wavelength, Waywardhorizons, WeißNix, Wereon, Will Beback, Wilsbadkarma, Wimt, Yellowtruck, Ylem, Yngvadottir, Zanimum, Zap Rowsdower, Zfr, Zzuuzz, Александър, आशिष भटनागर, 718 anonymous edits

Census *Source*: http://en.wikipedia.org/w/index.php?title=Census *Contributors*: 16@r, 2toise, 90 Auto, A. B., ASDFGH, Abhaga, Achim Jäger, Adam Bishop, Aewold, AgainErick, Ahmedrefat, Ahmet Huner, Ahuskay, Aitias, Alansohn, Alex.muller, Alexd321, Alexzr88, Allstarecho, Altenmann, Andres68, Andrewpmk, Andy Marchbanks, Andycjp, Angrytoast, Angusmclellan, AnonMoos, Antiuser, Arcadian, Argon233, Arkaig1, Arnab1996, Artaxiad, Ashley thomas80, Australia boy, Avala, Avenged Eightfold, Avoided, BD2412, Badanedwa, BanyanTree, Barefeetdude, Barkers PR, Bart Versieck, Bender235, Benlisquare, Berean Hunter, Berkunt, Bersus, Bffjil, Bfswebstat, Bhadani, Bigsean0300, Birnuson, Biruitorul, Bitbut, Blackknight12, Blehimbored:), Bloodshedder, Bobblewik, Bogdangiusca, Bongwarrior, Boomer1234567, Boreas74, Boxplot, Brandmeister (old), Brianjd, Brutalhovno, Bry9000, Bucketsofg, CRICKETLtd, CTF83!, Ca3tki, Calliopejen1, Calvintiger, CanOfWorms, Canadaonline, Carboli, CaribDigita, Carwil, Cdc, Celtmist, Cgingold, Cherkash, Chester Markel, Chirag, ChrisGualtieri, ChrisNickson, Chrislk02, Chrism, Christian List, Chrono1084, Chuq, Ckatz, Closedmouth, Cmapm, Cnees, Colchicum, CommonsDelinker, Conversion script, Coopkev2, Cordless Larry, Courcelles, Cpl Syx, Crazyquesadilla, Cribbswh, Cstreet, Cup o' Java, Curb Chain, Cyanoa Crylate, Cymru.lass, D. F. Schmidt, DBigXray, DS1953, Dale Arnett, Dan653, DanielCD, Datu Dong, David Thrale, David1217, Davidcannon, Dbachmann, Dcastor, Ddroar, Deeptrivia, Delirium, Denisarona, Dewet, Dgkuty, Die spam die, Dinakarr, Dirkbb, Discospinster, Doggitydogs, Dominictroc, Donald Albury, Donfbreed, Dongjing, Doric Loon, Doseiai2, Dramatic, Drozmight, Dub8lad1, Dwight666, EamonnPKeane, Earl Andrew, Edcolins, Edward, Eilthireach, Eleassar, EmanWilm, Emilios1310, Emperorbma, EqualRights, Eregli bob, Erianna, Eric Wester, Erkan Yilmaz, Estoy Aquí, EugeneZelenko, Evil Monkey, Exec8, Ezhiki, Facial, FakTNeviM, Farhan Zafar, Fatstraw, Favonian, FedLawyer, Fern Forest, Fg2, Flapdragon, Flying Saucer, FocalPoint, Footballover, Fratrep, Frazzle, Fredbauder, Freshbakedpie, Futurist110, Fz22, Fæ, GABaker, Gaius Cornelius, Galoubet, Garrettmca, Gazilion, GenericBob, Genev82, Geoffrie, Gidonb, Giraffedata, Gogo Dodo, Gointemm, Gookhatingvet, Gordon.lunar, GraemeL, Graham87, Granucci, Grick, GringoInChile, Ground Zero, Guest9999, Gun Powder Ma, H2g2bob, H@r@ld, HJ Mitchell, Hadal, Hagedis, Hairy Dude, Halibutt, Hank7hill, Hattrem, HenryLi, Hephaestos, Heyhey Hey, Hibernian, Hmains, Hohum, Homerjay, Hut 8.5, Hwy43, Hydrogen Iodide, IRP, IRWolfie-, Igoldste, Ihatepotsmokinghippies, Iiiiaaaa, IkonicDeath, Imoen, IndepIntel, Intgr, Iridescent, Isaac120, Isthmus, Itsmejudith, JForget, JHunterJ, JackofOz, Jagged 85, Jannismakris, Jauerback, JavierMC, Jboyle4eva, Jeandré du Toit, Jeff G., Jeraphine Gryphon, Jerry Ritcey, Jerzy, Jibril23, Jim Huggins, Jim no.6, Jimp, Jiri007, Jm1122k, Jnestorius, Joao Xavier, JoeSmack, John ellenberger, John of Reading, JohnGurney, Johnjay1745, JonHarder, Jonemerson, Jonesey67, JorgeGG, JorgeRodriguez, Joseph Solis in Australia, Joseph simeon, Josette, Julius Sahara, Julo, Jusdafax, Jusjih, Just plain Bill, Just4ikarus, KGasso, Kaihsu, Kat263, Kay556690, Kayau, Keith Edkins, Kembangraps, Khalil babli, Kiefer.Wolfowitz, Kigali1, Kim bas, King Lopez, KingKane, Kingpin13, KirrVlad, Kku, Koroner, Krikwen, Kristaga, Kross, Kyky108, L Kensington, Lacrimosus, Laogeodritt, Lasunncty, Lateg, Lauzonda, Lawlessurchin, Ldsviking, Legumeman, Leszek Jańczuk, Levineps, Li-sung, Licor, Lindar, LizardJr8, LlywelynII, Llywrch, Logicwiki, Ltool63, Luna Santin, Luntertun, LynxNukem, Mack2, MadGeographer, Magbatz, Malcolm Farmer, Mandarax, Mandsford, Manuel Anastácio, Marathone, Marek69, Mark Arsten, Martpol, Masoudfcr, Materialscientist, Matt Brennen, Matthead, Mattsains, McSly, Meiskam, Meister, Melcombe, Mentifisto, Mhym, Michael Devore, Michael Hardy, MichaelCarl, Midway, Mikkel, Milboz, Mild Bill Hiccup, MisfitToys, Misskaykay, MithrandirAgain, Mkanoni, Modulatum, Mr.Badlands, MrBell, MrFawwaz, Mrh30, Muppet, Mxn, MysticalCow, Mzajac, Mátyás, Nabob Nagusa, Nachoman-au, Nateofgreat, Nauseam, NawlinWiki, NeonMerlin, NerdyScienceDude, Nestled like spoons, Neurolysis, Newhoggy, Nfitz, Nhgenealogy, Ninsheart, Niteowlneils, Nmacu, Nn123645, Nomi887, Norbu19, Notheruser, Nponte, Nrlight, Ns.code, NuclearWarfare, Numbo3, Nunquam Dormio, Nyttend, O'Dea, Ohconfucius, Oleg Alexandrov, Oliver Pereira, Olyus, Omdfw, One, Onya4ever, Optimist on the run, Orodreth, PANONIAN, PMLF, PRODUCER, Panapp, PaoloValente, Parishan, PatrickFlaherty, Paul Norfolk Dumpling, Paul W, Paul-L, Pepith, Pgallert, Phantomsteve, Pharaoh of the Wizards, Philip Trueman, Philopp, Pigman, Pinethicket, Piotrus, Plastikspork, Pne, PoeticVerse, Polly Math, Poor Yorick, Priyansh verma, Prop3v56, Quadell, Quadparty, Quebec99, Queenmomcat, Quicksilvre, Qwertzy, Qwfp, RODERICKMOLASAR, RTG, Rami R, Rammy2cool, Razor2988, Reach Out to the Truth, Reaper Eternal, Red Slash, Reinyday, Remigiu, Ren Sydrick, RenamedUser01302013, Rentzepopoulos, Rhopkins8, Rich Farmbrough, Richard Arthur Norton (1958-), Rjwilmsi, Rmhermen, Rollon, Romanm, Ronhjones, Rui Gabriel Correia, Ryl411, Saddhiyama, Sahasrahla, Sandstein, Sannse, SatyrTN, Savidan, Scientus, ScottyBerg, Sdaeun, Seaphoto, Selket, SemperBlotto, Senator2029, Sesu Prime, Sfs90, Shadowjams, Shannon bohle, Shii, Shoessss, Siamax, Sift&Winnow, Simetrical, Sionk, Sivuman, Sjakkalle, Skizzik, SlimshadY, Smallman12q, Smarkflea, Smooth O, So-nj, Sombrero, Sortcoach, SpacemanSpork, Spearhead, Specs112, Spoladore, Srich32977, StAnselm, StaticGull, Stet, Sukarnobhumibol, Tarih, Tarun mor50002, Taztouz, Tempac, Terovian, The Rambling Man, The Thing That Should Not Be, TheRingess, Thunderbuck ram, Tide rolls, Tim1357, Timc, Timneu22, Timrollpickering, Tomcakes, Tomi, Tommy2010, Tomsega, Tony Corsini, Tpbradbury, Treisijs, Triona, Truthanado, Tualha, Twhyatt, Typing General, UberScienceNerd, Ultraboy8888, Uncle Dick, Unnikn, Urg writer, Utar, Valhalla, Vanished user 342562, Vicki Rosenzweig, VictorianMutant, Wavelength, Wayne Riddock, Wheatleya, WhiteWriter, Whodhellknew, Wienertakesall, Wiki-uk, Wiki1609, WikiPuppies, Wikidan829, Will Beback Auto, Wimt, Wolfvuk, Wtshymanski, Wysprgr2005, XNanakate, Xanthar, Ycu1101ycu, Yidisheryid, Yngvadottir, Ynhockey, Yom, Yoman82, YuMaNuMa, Yupik, Zaian, Zangar, Zigger, Zinc2005, Zlerman, Zollerriia, Zundark, Zxcvn80, Zzuuzz, نگونبانگونی, اشف عقیل, 847 anonymous edits

List_of_highest-income_places_in_the_United_States *Source*: http://en.wikipedia.org/w/index.php?title=List_of_highest-income_places_in_the_United_States *Contributors*: 2A02:2F09:607F:FFFF:0:0:4F77:A615, 72Dino, AED, Alansohn, Alcea setosa, Angelo De La Paz, AniMate, Anonymous26, AntonioJackson, Aoa8212, Ardis12, Aslayerfan, Astuishin, Auntof6, Bcorr, Bdesham, Bkonrad, Bmzhu, BrendelSignature, Bud08, Canderson7, Cburnett, Cg41386, Cherkash, Christopher Parham, Cjcamarillo, Coulraphobic123, D6, DPCBOSS, Dolexpo,

Dpbsmith, DrJayDee, ES80829, Eastlaw, Elockid, Esaontheroad, Fenix down, Flubber12345, Freddiem, Funandtrvl, Geni, Gsdlova, Hertz1888, Hmains, Hrearden, Hydrogen Iodide, JDG1992, JVittes, Jajhill, Jameswilson, Jared Kofsky, Jdaloner, Juicifer, Ken Gallager, Lakester10, LawerenceSmit, Lhakthong, Longhorns4life89, Lukaniko, Marcus Qwertyus, Martiniano, Michael Hardy, Michaelenandry, Mike Dillon, Mwalcoff, Njsuplex, Nyttend, Old Guard, PDH, Paperfire, Peace and Passion, RA0808, Ravedave, Reinyday, Retired username, Reywas92, Rich Farmbrough, Ricky81682, Sbmeirow, Seven Days, Sfahey, Shamir1, Short Verses, ShortySQuicker, Skatman, Sonicboi, Splash, Supermage, Sverdrup, Tdl1060, Tgeairn, Thayer, Thecheesykid, Tide rolls, Tombo7791, TravisTX, Tszhangto, UnitedStatesian, WayneMokane, Weedwhacker11, Whyaduck, Wikiman8531212, Will Beback, Zalgo, 513 anonymous edits

Poverty_threshold *Source*: http://en.wikipedia.org/w/index.php?title=Poverty_threshold *Contributors*: Aaboltaabol, Acroterion, Aengus, AeriDenise, Agradman, Ahackworth, Aillema, Aleksd, Aliyevramin, All Is One, Allens, Amazingcreator79, Andy Xucha, Anomalocaris, Antandrus, Arctic Kangaroo, Arthena, Astronautics, AuburnPilot, Ausir, Barek, Beland, Bigpapi6292, Bluerasberry, Bookandcoffee, BrendelSignature, Burrburr, Byelf2007, CLW, Can't sleep, clown will eat me, Cantus, Cardsplayer4life, Cassivs, Cconnett, Cflm001, Chozanm, Chuckdevore, Cielomobile, Cirino73, Ckatz, Classicrockfan42, Closenplay, Cntras, Cocoaguy, Coelhodg, Commander Keane, Common Man, Cretog8, Cuvtixo, Dambrosio, Damicatz, DanKeshet, David Biddulph, DevilEditerLol, Dgies, Dismas, Dlohcierekim, Drmies, Duoduoduo, Dwrcan, Dystopos, Ed Poor, Edgar181, Edward, Egoadvocate, ElentariAchaea, ElinorD, Elitehobo, Epbr123, Eraserhead1, Eurosong, Fakelvis, Falconleaf, Fasttwitch, Fconaway, Fidalgo, Fitzaubrey, Foober, Frenzie, Gaff, Galf, Gareth Griffith-Jones, Gavin.perch, Gene.arboit, Gilliam, Gizziiusa, Globalsolidarity, Goldom, Grick, Guoguo12, Hadrianheugh, Hamedy1986, Haoie, HarveyA, Hector.C.Jorge, Hede2000, Henrygb, Heron, HexaChord, Hibernian, Hrh80, Human.v2.0, Humanflux, Hurutch, Ian13, Ivan1984, Iviney, J.delanoy, JakubATusaDotcom, Janshouse, JavaAndCSharp, Javadane, JeremyA, Jhortman, Jim1138, Jimtaip, Jitse Niesen, Joseph Solis in Australia, Josh Parris, Joy, Junling, Kanook, KathrynLybarger, Kellen, Kelly Martin, Ken@popp2.com, Khanra, Kirachinmoku, Kissmequick, KungFuMonkey, Lalit Jagannath, Lawrencekhoo, Leslie Mateus, Leszek Jańczuk, Levineps, LilHelpa, LilyKitty, Logical Fuzz, LolChavista, Lost tourist, MOA12345, Maralia, Marek69, Materialscientist, Mattbrundage, Matthewcgirling, Maurreen, Maxis ftw, McDutchie, Metta Bubble, Million Moments, Minimac, Misanthr0pe, Monsday, Mrshannigan, Mshecket, Muhandes, Musiphil, Mydogategodshat, Mynameisnotpj, Mysdaao, Möchtegern, Nach0king, Nagy, Naniwako, Naval Scene, Neptunius, Niceguyedc, NigelR, Nihiltres, Nishkid64, Noodleki, Noozgroop, Oceanflynn, Oledr, Otolemur crassicaudatus, P.bertazzi, Paddles, Passionless, Philip Trueman, Phoebe, Piotrus, Policy Person, Poopeater111, Postmortemjapan, Professortimithy, Quest for Truth, QuiteUnusual, R1000R1000, RJ010890, RainbowOfLight, ReiVaX, Revth, RexNL, Riana, Rich Farmbrough, Rjwilmsi, RobyWayne, Roke, SDnogo, Sam Hocevar, Sammo, Sapphic, Sarahj2107, Scarsxrunxdeep, SeanAhern, Search4Lancer, Septegram, Sh-otaibi, Shannon bohle, Sharkface217, Sharnak, Sheetal.sehgal, Shoy, SiobhanHansa, Skysmith, Slhurst, Smallman12q, Smokecrack, Snipson23, Squashy, Sreenand wiki, Staszek Lem, Strait, Tanaats, Team9a, Terjepetersen, The Mark of the Beast, The man stephen, Thingg, Tigga, Timc, Timwi, Tinton5, Toh, Tommy2010, TouristPhilosopher, Twang, Ultramarine, Valdomiro, Vanished user uih38riiw4hjlsd, Virtakuono, Volunteer Marek, Vungadu, WAGATHEE, WhatamIdoing, Wing Nut, Woutersmet, Wtmitchell, Yaktx, Yoyoyoitsrob, Zappa711, Zouf, 378 anonymous edits

Time_zone *Source*: http://en.wikipedia.org/w/index.php?title=Time_zone *Contributors*: *drew, 159753, A. B., A.Ou, A0404141041, A8UDI, AHeneen, ANDREJACK, Aaaallleex, Aaronw, Absalom89, Acegikmo1, Acs4b, AfDproXX, Ahoerstemeier, Alanmak, Alansohn, Albmont, Alex Bakharev, Alksentrs, AllyUnion, Alynna Kasmira, Ambarish, Aminullah, Amit6, Amolkold, Amorim Parga, Amorymeltzer, Andre Engels, Andrewpmk, Androl, Anneman, Antandrus, Arichnad, Arthena, Ashley Y, Asterion, Australia boy, AxelBoldt, BD2412, BIL, Banhtrung1, Basil.bourque, Bastin, Bazonka, Bdodo1992, Bearcat, Beland, Bellczar, Ben Arnold, Bentogoa, Bermlee, Bethpage89, Big Adamsky, Billybob112, Bjankuloski06en, Bjarki S, Bjcairns, Blahma, Bluelion, Bobo192, Bomac, Bongwarrior, Breno, Brian Kendig, Brian Patrie, Bryan Derksen, Budlight13, Buggery Boar, Bunthorne, Bushcarrot, CQ, Caese, Calor, Calton, CambridgeBayWeather, Can't sleep, clown will eat me, Canjth, CaribDigita, CatherineMunro, Celarnor, Cenarium, Centpacrr, CesarB, Charlene.fic, Chihuahua State, Chipmunkdavis, Chmod007, Chochopk, Chris24, ChrisGualtieri, Christy747, Chuck SMITH, Chun-hian, Chuq, Chuunen Baka, Circeus, Cirt, Classical geographer, Cliffb, Closedmouth, Cnmuk, Coder Dan, ColinKennedy, Colonel Cow, Commdor, CommonsDelinker, Comrade Graham, Confer44, Conscious, Conversion script, CopperSquare, Corumplex, Crd721, Cst17, Curps, Cyrus the virus grissimo, D-Rock, D15724C710N, DARTH SIDIOUS 2, DMarvin, DO'Neil, DPH1110, DVD R W, Da Vynci, Dale Arnett, Damian Yerrick, Damirgraffiti, Darth Panda, David Biddulph, DavidGrayson, Davidcannon, Davodd, Dcsohl, Deelkar, Denelson83, Derek Ross, Derild4921, Disill, Diti, Djszucs, Dmacgr 22, Dmeranda, Dmh, Dmn, DocWatson42, Docu, Donarreiskoffer, Doniago, Donner60, DoubleBlue, DouglasCalvert, Download, DrBizarro, Dragon taker, Drbreznjev, Drbug, Dreadstar, Drh104, Dsyzdek, Duncan, Duncan7670, DyNama, Dysepsion, E. Ripley, ESkog, EamonnPKeane, Earl Andrew, Earthlyreason, Ebelular, Ed Poor, Edward, Eequor, Eerlend, Ehrenkater, Ekren, Enchanter, Eomund, Eoujw, Epbr123, Eraserhead1, Erasmussen, Eric Herriman, Eric-Wester, Eric119, Esmito, Esperant, Eubulides, Euchiasmus, Evice, Evil Monkey, Evil saltine, Ewawer, Excirial, Ezhiki, Ezhuttukari, F. F. Fjodor, FB 2000, FF2010, Facts707, Fanyavizuri, Farosdaughter, Farzanegan, FayssalF, Federico Tortorelli, Feedmecereal, Feezo, Feinoha, Fennec, Ferkelparade, Fffdervis, Fieldday-sunday, Fmph, Fonadier, Forseti, FrancoGG, Freddiecaveman, FreplySpang, Fresheneesz, Funandtrvl, Fundamental metric tensor, Fuzheado, Gabbe, Ganteng, Garfield226, Gary King, Gerry Ashton, Giant89, Giftlite, Glane23, Glenn, Glenn L, Gogo Dodo, Golbez, Goobergunch, GorillaWarfare, Gorrister, Goudzovski, Graeme Bartlett, Grafen, Graham87, Grassfire, Grick, Groogle, Ground Zero, Grstain, Guaka, Gzornenplatz, H2g2bob, HJ Mitchell, Hadal, Hadiyazzi, Haham hanuka, Haihe, Haikupoet, Hajor, Happening1, Hapsiainen, Harp, Harriv, Hazhk, Heb, Heitordp, Hellerick, Henrygb, Hesperian, Hi is bob, HistoryAddict, HortusBlizzard, Htmlland, Humilulo, HumphreyW, Hvn0413, Hydrargyrum, IKidaround, IMSoP, IQveteran, Ido50, Ignis Fatuus, IgorMagic, Imoen, Imz, Incnis Mrsi, Indefatigable, Iner22, IngSoc BigBrother, Instantnood, Intgr, Iridescent, Island, Itai, Itisonly4me, Ixfd64, J-stan, J.delanoy, J.smith, JHunterJ, JLaTondre, JNW, JackofOz, Jakes18, Jamadagni, Jarrodiscool141314, JavierMC, Jc3s5h, Jcra, Jd4508, Jda, Jeff G., Jeff3000, Jeffq, Jerome Charles Potts, Jerzy, Jesper Norgaard, Jiang, Jim1138, Jimp, Jj137, Jla, Jnestorius, Joe Kress, JoeSmack, Joey2cool2007, John Paul Parks, John Quincy Adding Machine, John of Reading, JohnOwens, JohnWittle, Jojhutton, Jojit fb, JonathanJosephsen, Jonathunder, Joostjodel, Joseph Solis in Australia, Jpk, Jplimpz, JulDes, Juliancolton, Julianp, Jumbuck, Jusjih, Just H, Jérôme, Karl Palmen, Karl-Henner, KathrynLybarger, KayEss, Ke4roh, Keeves, KeithH, Kh2mohamed, King of Hearts, Kingal86, Kiwi128, Klemen Kocjancic, Knowledge Seeker, Krawunsel, Kubigula, Kukini, Kungfuadam, Kyle Barbour, Kylekieran, LOL, Lacrimosus, Lalah1294, Lambiam, Lambyte, Lamrongpeng5, Last Avenue, Lee J Haywood, LeeTraynor, Leeyc0, LeinsterLion, Leucius, Leuko, Lewisjon, LilHelpa, Lincmad, Lir, Lizzie Harrison, Llywrch, Logan, Lolm8, LorenzoB, Lotje, Lucamauri, Lugia2453, Luis Dantas, Luna Santin, M, MKar, Mac, MacTire02, Macaddct1984, Magnus Manske, Maksim L., Malrase, Mancunius, Marechal Ney, Marek69, Mariusmiti, Mark Leander, MarkSweep, Marshman, Martarius, Mathew12123434, MatthewVanitas, Mav, MaxDZ8, Mayfoev, Mayooresan, McSly, Mean as custard, Meelar, Meini, MelbourneStar, Mernen, Mert34, Meters, Michael Glass, Michael Hardy, MichaelBillington, Michfan2123, Midnightblueowl, Mild Bill Hiccup, Minipie8, Minna Sora no Shita, Miquonranger03, Miranda, Miros 0571, Mjs1991, Modulatum, Mojo-chan, Monkeyman, Monni95, Monoxide333, Montrealais, Moocha, MoraSique, Morga555, Mormegil, Mr Bartels, Mr. Ajax, MrWolfy, Mrk695, Mryg, Mwarren us, Mxn, N1ywb, Naive cynic, Nanshu, Natpowning, Naval Scene, NawlinWiki, Nealmcb, NellieBly, Neptune777, NewEnglandYankee, Nichalp, Nicholas Weiner, Nicolas Love, Nightstallion, Niteowlneils, Nitsansh, Nk, Northumbrian, Nortonius, Novalis, Numbo3, Numen, Nunh-huh, Octahedron80, Ohnoitsjamie, Oleksiy.n, Olivier, OpenFuture, Optimist on the run, OverlordQ, OwenX, Oxymoron83, P. S. F. Freitas, PDucks, Paddu, Pak21, Panos84, Paranoid, Pasquale, Patach, Patcs0013, Patrick, Paul August, Paul Koning, Paul Murray, Paullees, Pavel Vozenilek, Pejman47, Peregrine981, Persian Poet Gal, Peter Wisner, PeterThoeny, Petrig, Pgan002, Phearson, Philip Stevens, Philip Trueman, Phoenix B 1of3, Pi zero, Piano non troppo, Picapica, Picaroon, PierreAbbat, PimRijkee, Pm5, Pokrajac, Popsracer, Possum, Preslethe, Psb777, Psdekloe, Pseudonymous Rex, Pytom, Quadell, Quasipalm, Rabbit233224, Radiojon, RainbowOfLight, Randyc, Raymondwinn, Rchandra, Rclocher3, Reaper Eternal, Redrose64, Redwood57, Reisio, Renetus, RenniePet, Requestion, RexNL, Rhobite, Riana, Rich Farmbrough, Rjwilmsi, Rmallins, Rmhermen, Robsinden, Roland2, Romanm, Rory096, Royaume du Maroc, Rsrikanth05, Rt66lt, Russell E, S.Örvarr.S, SGBailey, SL93, SV Resolution, Sahgal, Samuel, Samw, Saros136, Sceptre, SchmuckyTheCat, Schneelocke, Schoen, Scipius, Scoelho86, Scott Roy Atwood, Scott3, Seba5618, Secfan, SeoMac, SephChicago, Sephiroth BCR, Sesu Prime, Seth Ilys, Sewing, Sheridan A.M. Hamilton, Shoefly, Sietse Snel, Simon J Kissane, Simon_J_Kissane, Simpsnut14, Siqbal, Sjorford, Skater, Skizzik, Skyman123, Skysmurf, Skywayman, Slawojarek, Smalljim, Smart77, Smithcraft36, Smjg, Smoken Flames, Snowdog, Sommerfeld, Sonninkikkeli, SouthernNights, Sp33dyphil, Spesh531, Spitfire, Squiffycat, Steel, Stephan Leeds, Stephen B Streater, Stephen.frede, SteveMX, Sunborn, Susurrus, Sverdrup, Swid, Swpb, T-resh, T.c.w7468, TGC55, TPK, TUF-KAT, TZ master, Tad Lincoln, Takita787, TakuyaMurata, Tarquin, Tashikelzang, Tbhotch, Tebloraf, Techman224, Techtonik, Teemu Leisti, Telempe, Tfts, That-Vela-Fella, The Duke of Waltham, The High Fin Sperm Whale, The Rambling Man, The Storm Surfer, The Thing That Should Not Be, The Tom, TheEgyptian, Thf1977, Thingg, Think outside the box, Thryduulf, Thumperward, Tide rolls, Tim Parenti, Time in Russia, TimeCurrency, TimeXCode, TimeZonesBoy, Timir2, Timneu22, Timwi, Tivolives, Tmangray, Tmol42, Tobias Conradi, Tomica, Topdeck, TopoChecker, Topory, Torpov, Transisto, TruPepitoM, TrueMahn, Tsja, Tutsuro, UhMaZe xpeK, Ultratomio, Uncle Scrooge, Undescribed, Uriber, Ute in DC, VampWillow, Vanished User 1004, Vanjagenije, Vaughan Pratt, Vee power, Velho, Versus22, Viriditas, Vsmith, Walshie79, Wangi, Watson24hour, Wavelength, WeißNix, Weregerbil, West.andrew.g, Wexcan, Widr, Wik, Wiki Wikardo, Wikidenizen, Wikilackey, Wikipediman, Will Beback Auto, Willking1979, Wissons, Wizard04, Woodstone, Woohookitty, Worldtimezone, Writ Keeper, Writherthebest100, Wrtiede, Wtmitchell, Wwutx, Xeryus, Ximi12, Xpi6, Xxeric thomasxx, Xxxx693, Xzqx, Y0u, YUL89YYZ, Yamara, Ybbor, YellowMonkey, Yosef1987, Yt788, Zakuragi, Zash20, Zidonuke, Zkzkz, Zoicon5, ZooFari, Zundark, Zzuuzz, Σ, Владимир Шеляпин, תלחק, 1556 anonymous edits

Image Sources, Licenses and Contributors

File:Plaque on the statue next to the Frick Fine Arts Building.JPG *Source*: http://en.wikipedia.org/w/index.php?title=File:Plaque_on_the_statue_next_to_the_Frick_Fine_Arts_Building.JPG *License*: unknown *Contributors*: User:Piotrus

File:AlleghenyCtyCourthouse-082904.jpg *Source*: http://en.wikipedia.org/w/index.php?title=File:AlleghenyCtyCourthouse-082904.jpg *License*: unknown *Contributors*: User:Bobak

File:Allegheny County Medical Examiner jeh.jpg *Source*: http://en.wikipedia.org/w/index.php?title=File:Allegheny_County_Medical_Examiner_jeh.jpg *License*: unknown *Contributors*: User:Jim.henderson

File:Map of Allegheny County Pennsylvania With Municipal and Township Labels.png *Source*: http://en.wikipedia.org/w/index.php?title=File:Map_of_Allegheny_County_Pennsylvania_With_Municipal_and_Township_Labels.png *License*: unknown *Contributors*: US Census, Ruhrfisch

File:Map of Allegheny County Pennsylvania School Districts.png *Source*: http://en.wikipedia.org/w/index.php?title=File:Map_of_Allegheny_County_Pennsylvania_School_Districts.png *License*: unknown *Contributors*: US Census, Ruhrfisch

File:I-79 (PA).svg *Source*: http://en.wikipedia.org/w/index.php?title=File:I-79_(PA).svg *License*: unknown *Contributors*: User:Jeff02

File:I-279 (PA).svg *Source*: http://en.wikipedia.org/w/index.php?title=File:I-279_(PA).svg *License*: unknown *Contributors*: User:Jeff02

File:I-579 (PA).svg *Source*: http://en.wikipedia.org/w/index.php?title=File:I-579_(PA).svg *License*: unknown *Contributors*: User:Jeff02

File:I-76 (PA).svg *Source*: http://en.wikipedia.org/w/index.php?title=File:I-76_(PA).svg *License*: unknown *Contributors*: User:Jeff02

File:Pennsylvania Turnpike logo.svg *Source*: http://en.wikipedia.org/w/index.php?title=File:Pennsylvania_Turnpike_logo.svg *License*: unknown *Contributors*: Dough4872, Dream out loud, NE2, O

File:I-376 (PA).svg *Source*: http://en.wikipedia.org/w/index.php?title=File:I-376_(PA).svg *License*: unknown *Contributors*: Ltljltlj

File:Turnpike-576.svg *Source*: http://en.wikipedia.org/w/index.php?title=File:Turnpike-576.svg *License*: unknown *Contributors*: Vishwin60

File:US 19.svg *Source*: http://en.wikipedia.org/w/index.php?title=File:US_19.svg *License*: unknown *Contributors*: Rocket000, SPUI, TwinsMetsFan, Washuotaku, 1 anonymous edits

File:US 22.svg *Source*: http://en.wikipedia.org/w/index.php?title=File:US_22.svg *License*: unknown *Contributors*: SPUI, 1 anonymous edits

File:US 30.svg *Source*: http://en.wikipedia.org/w/index.php?title=File:US_30.svg *License*: unknown *Contributors*: Allen4names, Bidgee, SPUI, TwinsMetsFan, 4 anonymous edits

file:speakerlink-new.svg *Source*: http://en.wikipedia.org/w/index.php?title=File:Speakerlink-new.svg *License*: unknown *Contributors*: User:Kelvinsong

File:Pennsylvania-counties-map.gif *Source*: http://en.wikipedia.org/w/index.php?title=File:Pennsylvania-counties-map.gif *License*: unknown *Contributors*: AnonMoos, Carnildo, JosN, Padraic, 2 anonymous edits

File:National-atlas-pennsylvania.png *Source*: http://en.wikipedia.org/w/index.php?title=File:National-atlas-pennsylvania.png *License*: unknown *Contributors*: Shauni

File:Edward Hicks - Penn's Treaty.jpeg *Source*: http://en.wikipedia.org/w/index.php?title=File:Edward_Hicks_-_Penn's_Treaty.jpeg *License*: unknown *Contributors*: Cobalty, Man vyi, Uyvsdi, Zolo, 1 anonymous edits

File:John Dickinson portrait.jpg *Source*: http://en.wikipedia.org/w/index.php?title=File:John_Dickinson_portrait.jpg *License*: unknown *Contributors*: User:PKM

File:PhiladelphiaPresidentsHouse.jpg *Source*: http://en.wikipedia.org/w/index.php?title=File:PhiladelphiaPresidentsHouse.jpg *License*: unknown *Contributors*: Aavindraa, BoringHistoryGuy, 1 anonymous edits

File:HillsCapitol.jpg *Source*: http://en.wikipedia.org/w/index.php?title=File:HillsCapitol.jpg *License*: unknown *Contributors*: Original uploader was ClairSamoht at en.wikipedia

File:Federal Emergency Relief Administration, FERA camp for unemployed women in Arcola, PA - NARA - 196583.tif *Source*: http://en.wikipedia.org/w/index.php?title=File:Federal_Emergency_Relief_Administration,_FERA_camp_for_unemployed_women_in_Arcola,_PA_-_NARA_-_196583.tif *License*: unknown *Contributors*: Howcheng, Infrogmation

Image:PennsylvaniaDemography.svg *Source*: http://en.wikipedia.org/w/index.php?title=File:PennsylvaniaDemography.svg *License*: unknown *Contributors*: User:Quilokos

File:Pennsylvania population map 1.png *Source*: http://en.wikipedia.org/w/index.php?title=File:Pennsylvania_population_map_1.png *License*: unknown *Contributors*: User:JimIrwin

File:Bethlehem Steel.jpg *Source*: http://en.wikipedia.org/w/index.php?title=File:Bethlehem_Steel.jpg *License*: unknown *Contributors*: Original uploader was Jschnalzer at en.wikipedia

Image:Crayola-64.jpg *Source*: http://en.wikipedia.org/w/index.php?title=File:Crayola-64.jpg *License*: unknown *Contributors*: Dougie WII

File:Philly skyline.jpg *Source*: http://en.wikipedia.org/w/index.php?title=File:Philly_skyline.jpg *License*: unknown *Contributors*: Ed Yakovich , emy111 , flickr: i like the weekend , http://www.flickr.com/photos/10396190@N04/. Original uploader was Emy111 at en.wikipedia

File:Pittsburgh WEO Night 1.jpg *Source*: http://en.wikipedia.org/w/index.php?title=File:Pittsburgh_WEO_Night_1.jpg *License*: unknown *Contributors*: User:ronjamin

File:Allentown.jpg *Source*: http://en.wikipedia.org/w/index.php?title=File:Allentown.jpg *License*: unknown *Contributors*: Paul Leiby

File:Penn campus 2.jpg *Source*: http://en.wikipedia.org/w/index.php?title=File:Penn_campus_2.jpg *License*: unknown *Contributors*: Bryan Y.W. Shin

Image:Dorney Park Steel Force Thunderhawk.jpg *Source*: http://en.wikipedia.org/w/index.php?title=File:Dorney_Park_Steel_Force_Thunderhawk.jpg *License*: unknown *Contributors*: Themeparkgc

File:Baughman's Rock Overlook.ogv *Source*: http://en.wikipedia.org/w/index.php?title=File:Baughman's_Rock_Overlook.ogv *License*: unknown *Contributors*: David Fulmer from Pittsburgh

Image:Allegheny Mountain Tunnel.JPG *Source*: http://en.wikipedia.org/w/index.php?title=File:Allegheny_Mountain_Tunnel.JPG *License*: unknown *Contributors*: Taken by Willkie Cirker. User Bcirker on en.wikipedia

Image:2215 - Pennsylvania Turnpike Tunnel.JPG *Source*: http://en.wikipedia.org/w/index.php?title=File:2215_-_Pennsylvania_Turnpike_Tunnel.JPG *License*: unknown *Contributors*: User:Thisisbossi

File:Appalachian Throughway.jpg *Source*: http://en.wikipedia.org/w/index.php?title=File:Appalachian_Throughway.jpg *License*: unknown *Contributors*: Nicholas

Image:Cokepark allentown.jpg *Source*: http://en.wikipedia.org/w/index.php?title=File:Cokepark_allentown.jpg *License*: unknown *Contributors*: User:Spikerogan

Image:Pocono Victory Lane.JPG *Source*: http://en.wikipedia.org/w/index.php?title=File:Pocono_Victory_Lane.JPG *License*: unknown *Contributors*: Morio, STBR

File:RuffedGrouse23.jpg *Source*: http://en.wikipedia.org/w/index.php?title=File:RuffedGrouse23.jpg *License*: unknown *Contributors*: Unknown/Uncredited - assumed USFWS employee.. Original uploader was Big iron at en.wikipedia

File:USBrigNiagaraInPort.JPG *Source*: http://en.wikipedia.org/w/index.php?title=File:USBrigNiagaraInPort.JPG *License*: unknown *Contributors*: User:Howcheng

File:Downtown Aliquippa.jpg *Source*: http://en.wikipedia.org/w/index.php?title=File:Downtown_Aliquippa.jpg *License*: unknown *Contributors*: User:Mvincec

File:AllentownPA Skyline.jpg *Source*: http://en.wikipedia.org/w/index.php?title=File:AllentownPA_Skyline.jpg *License*: unknown *Contributors*: Thomas Grim - twg1942

File:Altoona Downtown from Brush Mountain.jpg *Source*: http://en.wikipedia.org/w/index.php?title=File:Altoona_Downtown_from_Brush_Mountain.jpg *License*: unknown *Contributors*: User:Csoldner187

File:Beaver Falls.jpg *Source*: http://en.wikipedia.org/w/index.php?title=File:Beaver_Falls.jpg *License*: unknown *Contributors*: MrHope, Nyttend

File:Bedford pitt street.jpg *Source*: http://en.wikipedia.org/w/index.php?title=File:Bedford_pitt_street.jpg *License*: unknown *Contributors*: User:AudeVivere

File:Bethlehem Pennsylvania downtown.jpg *Source*: http://en.wikipedia.org/w/index.php?title=File:Bethlehem_Pennsylvania_downtown.jpg *License*: unknown *Contributors*: Tim Kiser ()

File:Blairsville Pennsylvania Market Street 2010.jpg *Source*: http://en.wikipedia.org/w/index.php?title=File:Blairsville_Pennsylvania_Market_Street_2010.jpg *License*: unknown *Contributors*: User:Canadian2006

File:Butler PA skyline.jpg *Source*: http://en.wikipedia.org/w/index.php?title=File:Butler_PA_skyline.jpg *License*: unknown *Contributors*: Mvincec. Original uploader was Mvincec at en.wikipedia

File:Carlisle, Pennsylvania.jpg *Source*: http://en.wikipedia.org/w/index.php?title=File:Carlisle,_Pennsylvania.jpg *License*: unknown *Contributors*: Doug Kerr

File:Memorial Square Chambersburg.jpg *Source*: http://en.wikipedia.org/w/index.php?title=File:Memorial_Square_Chambersburg.jpg *License*: unknown *Contributors*: Epstein.Mark, Euchiasmus, Jmabel, Vpuliva, 1 anonymous edits

File:Clarion, Pennsylvania.jpg *Source*: http://en.wikipedia.org/w/index.php?title=File:Clarion,_Pennsylvania.jpg *License*: unknown *Contributors*: Doug Kerr

File:Clearfield County Courthouse Apr 10.JPG *Source*: http://en.wikipedia.org/w/index.php?title=File:Clearfield_County_Courthouse_Apr_10.JPG *License*: unknown *Contributors*: Mackensen

File:Corry, Pennsylvania Skyline.jpg *Source*: http://en.wikipedia.org/w/index.php?title=File:Corry,_Pennsylvania_Skyline.jpg *License*: unknown *Contributors*: Nicholas

File:Pottercourthousecoudersport.jpg *Source*: http://en.wikipedia.org/w/index.php?title=File:Pottercourthousecoudersport.jpg *License*: unknown *Contributors*: Choess, Mackensen

File:Easton Skyline.jpg *Source*: http://en.wikipedia.org/w/index.php?title=File:Easton_Skyline.jpg *License*: unknown *Contributors*: Original uploader was Masesprint at en.wikipedia

File:Cambria County Courthouse - Ebensburg, PA.jpg *Source*: http://en.wikipedia.org/w/index.php?title=File:Cambria_County_Courthouse_-_Ebensburg,_PA.jpg *License*: unknown *Contributors*: User:Publichall

File:edinboro.jpg *Source*: http://en.wikipedia.org/w/index.php?title=File:Edinboro.jpg *License*: unknown *Contributors*: John of Reading, Taurus9395

File:Eriesky2.jpg *Source*: http://en.wikipedia.org/w/index.php?title=File:Eriesky2.jpg *License*: unknown *Contributors*: Original uploader was Pnoble805 at en.wikipedia

File:Franklin, Pennsylvania.jpg *Source*: http://en.wikipedia.org/w/index.php?title=File:Franklin,_Pennsylvania.jpg *License*: unknown *Contributors*: Doug Kerr

File:DowntownGettysburgPA.jpg *Source*: http://en.wikipedia.org/w/index.php?title=File:DowntownGettysburgPA.jpg *License*: unknown *Contributors*: Aurora30

File:Greencastle PA Square.JPG *Source*: http://en.wikipedia.org/w/index.php?title=File:Greencastle_PA_Square.JPG *License*: unknown *Contributors*: User:Smallbones

File:Greensburgskyline.jpg *Source*: http://en.wikipedia.org/w/index.php?title=File:Greensburgskyline.jpg *License*: unknown *Contributors*: User:Neurodistortion

File:Hanover theatre PA.jpg *Source*: http://en.wikipedia.org/w/index.php?title=File:Hanover_theatre_PA.jpg *License*: unknown *Contributors*: JOHN LLOYD

File:Downtown hazleton pa.jpg *Source*: http://en.wikipedia.org/w/index.php?title=File:Downtown_hazleton_pa.jpg *License*: unknown *Contributors*: Fenderbob, 2 anonymous edits

File:Hershey, Pennsylvania.JPG *Source*: http://en.wikipedia.org/w/index.php?title=File:Hershey,_Pennsylvania.JPG *License*: unknown *Contributors*: Original uploader was Bronayur at en.wikipedia

File:2008 03 30 - Hollidaysburg - Blair St 2.JPG *Source*: http://en.wikipedia.org/w/index.php?title=File:2008_03_30_-_Hollidaysburg_-_Blair_St_2.JPG *License*: unknown *Contributors*: User:Thisisbossi

File:Indiana Theater Sign.jpg *Source*: http://en.wikipedia.org/w/index.php?title=File:Indiana_Theater_Sign.jpg *License*: unknown *Contributors*: User:Scsmatt

File:Jeannette-pennsylvania-downtown.jpg *Source*: http://en.wikipedia.org/w/index.php?title=File:Jeannette-pennsylvania-downtown.jpg *License*: unknown *Contributors*: User:Canadian2006

File:Johnstownview.jpg *Source*: http://en.wikipedia.org/w/index.php?title=File:Johnstownview.jpg *License*: unknown *Contributors*: Auntof6, Jllm06

File:Kittanning fromhill.jpg *Source*: http://en.wikipedia.org/w/index.php?title=File:Kittanning_fromhill.jpg *License*: unknown *Contributors*: All Is One, Cngodles, 1 anonymous edits

File:Lancaster Pennsylvania downtown.jpg *Source*: http://en.wikipedia.org/w/index.php?title=File:Lancaster_Pennsylvania_downtown.jpg *License*: unknown *Contributors*: Randolph Carney

File:Latrobe PA MainStreet.jpg *Source*: http://en.wikipedia.org/w/index.php?title=File:Latrobe_PA_MainStreet.jpg *License*: unknown *Contributors*: User:Ohnoitsjamie

File:LebanonBolognaDrop2008.jpg *Source*: http://en.wikipedia.org/w/index.php?title=File:LebanonBolognaDrop2008.jpg *License*: unknown *Contributors*: Hooahman

File:Lewisburg, Pennsylvania.jpg *Source*: http://en.wikipedia.org/w/index.php?title=File:Lewisburg,_Pennsylvania.jpg *License*: unknown *Contributors*: Doug Kerr

File:ltownsquare1.jpg *Source*: http://en.wikipedia.org/w/index.php?title=File:Ltownsquare1.jpg *License*: unknown *Contributors*: User:KATMAAN

File:Clinton County Pennsylvania Courthouse.JPG *Source*: http://en.wikipedia.org/w/index.php?title=File:Clinton_County_Pennsylvania_Courthouse.JPG *License*: unknown *Contributors*: User:Ruhrfisch

File:McKeesRocksPA.jpg *Source*: http://en.wikipedia.org/w/index.php?title=File:McKeesRocksPA.jpg *License*: unknown *Contributors*:

File:Meadville, Pennsylvania.jpg *Source*: http://en.wikipedia.org/w/index.php?title=File:Meadville,_Pennsylvania.jpg *License*: unknown *Contributors*: Doug Kerr

File:Monessen01.JPG *Source*: http://en.wikipedia.org/w/index.php?title=File:Monessen01.JPG *License*: unknown *Contributors*: User:Mlouns

File:Kennedy Square.jpg *Source*: http://en.wikipedia.org/w/index.php?title=File:Kennedy_Square.jpg *License*: unknown *Contributors*: Nyttend

File:New Hope and Ivyland.JPG *Source*: http://en.wikipedia.org/w/index.php?title=File:New_Hope_and_Ivyland.JPG *License*: unknown *Contributors*: User:Jared Kofsky

File:Palmyra Downtown.jpg *Source*: http://en.wikipedia.org/w/index.php?title=File:Palmyra_Downtown.jpg *License*: unknown *Contributors*: Jayu from Harrisburg, PA, U.S.A.

File:Philadelphia Panorama From Camden.JPG *Source*: http://en.wikipedia.org/w/index.php?title=File:Philadelphia_Panorama_From_Camden.JPG *License*: unknown *Contributors*: User:Parent5446

File:Phoenixville downtown.jpg *Source*: http://en.wikipedia.org/w/index.php?title=File:Phoenixville_downtown.jpg *License*: unknown *Contributors*: User:Squelle

File:Duquesne Incline from top.jpg *Source*: http://en.wikipedia.org/w/index.php?title=File:Duquesne_Incline_from_top.jpg *License*: unknown *Contributors*: User:Plastikspork

File:Pittston City.jpg *Source*: http://en.wikipedia.org/w/index.php?title=File:Pittston_City.jpg *License*: unknown *Contributors*: User:vasiliymeshko

File:Pottstown PA HighStreet.jpg *Source*: http://en.wikipedia.org/w/index.php?title=File:Pottstown_PA_HighStreet.jpg *License*: unknown *Contributors*: Willjay

File:Pottsville.jpg *Source*: http://en.wikipedia.org/w/index.php?title=File:Pottsville.jpg *License*: unknown *Contributors*: Twinmama (talk). Original text: by Angela Henderson.

File:Groundhogday2005.jpg *Source*: http://en.wikipedia.org/w/index.php?title=File:Groundhogday2005.jpg *License*: unknown *Contributors*: Evrik, FlickreviewR, GrapedApe, Guety, Quadell, 1 anonymous edits

File:Reading, Pennsylvani skyline.jpg *Source*: http://en.wikipedia.org/w/index.php?title=File:Reading,_Pennsylvani_skyline.jpg *License*: unknown *Contributors*: Jack

File:Renovo, Pennsylvania 14 Street.jpg *Source*: http://en.wikipedia.org/w/index.php?title=File:Renovo,_Pennsylvania_14_Street.jpg *License*: unknown *Contributors*: User:Ruhrfisch

File:Ridgway PA from Country Club Apr 10.JPG *Source*: http://en.wikipedia.org/w/index.php?title=File:Ridgway_PA_from_Country_Club_Apr_10.JPG *License*: unknown *Contributors*: Auntof6, GrapedApe, Leyo, Mjrmtg

File:DowntownScranton2003.jpg *Source*: http://en.wikipedia.org/w/index.php?title=File:DowntownScranton2003.jpg *License*: unknown *Contributors*: Don Liotta. Original uploader was Donaldliotta at en.wikipedia

File:Selinsgrove, Pennsylvania.jpg *Source*: http://en.wikipedia.org/w/index.php?title=File:Selinsgrove,_Pennsylvania.jpg *License*: unknown *Contributors*: Doug Kerr

File:Uprising3.JPG *Source*: http://en.wikipedia.org/w/index.php?title=File:Uprising3.JPG *License*: unknown *Contributors*: Hal Smith

File:ThecornerSC.JPG *Source*: http://en.wikipedia.org/w/index.php?title=File:ThecornerSC.JPG *License*: unknown *Contributors*: Rafaelgarcia, Thisisbossi, Thuresson

File:Sunbury, Pennsylvania.jpg *Source*: http://en.wikipedia.org/w/index.php?title=File:Sunbury,_Pennsylvania.jpg *License*: unknown *Contributors*: Doug Kerr

File:Tyrone Borough Historic District Apr 10.JPG *Source*: http://en.wikipedia.org/w/index.php?title=File:Tyrone_Borough_Historic_District_Apr_10.JPG *License*: unknown *Contributors*: Auntof6, DanTD, Leyo

File:Uniontown PA.jpg *Source*: http://en.wikipedia.org/w/index.php?title=File:Uniontown_PA.jpg *License*: unknown *Contributors*: Guswrites, 2 anonymous edits

File:Warren Pennsylvania.jpg *Source*: http://en.wikipedia.org/w/index.php?title=File:Warren_Pennsylvania.jpg *License*: unknown *Contributors*: Nicholas from Pennsylvania, USA

File:WashingtonCountyCourthouse.JPG *Source*: http://en.wikipedia.org/w/index.php?title=File:WashingtonCountyCourthouse.JPG *License*: unknown *Contributors*: GrapedApe, Nyttend

File:Waynesboro, Pennsylvania.JPG *Source*: http://en.wikipedia.org/w/index.php?title=File:Waynesboro,_Pennsylvania.JPG *License*: unknown *Contributors*: User:Dincher

File:Wilkes Barre Downtown.jpg *Source*: http://en.wikipedia.org/w/index.php?title=File:Wilkes_Barre_Downtown.jpg *License*: unknown *Contributors*: User:vasiliymeshko

File:City Hall Williamsport Pennsylvania.JPG *Source*: http://en.wikipedia.org/w/index.php?title=File:City_Hall_Williamsport_Pennsylvania.JPG *License*: unknown *Contributors*: User:Ruhrfisch

File:Golden Plow York PA.JPG *Source*: http://en.wikipedia.org/w/index.php?title=File:Golden_Plow_York_PA.JPG *License*: unknown *Contributors*: User:Smallbones

Image:Countries by population density.svg *Source*: http://en.wikipedia.org/w/index.php?title=File:Countries_by_population_density.svg *License*: unknown *Contributors*: Miguel Contreras, Guatemala

File:World_population_density_1994.png *Source*: http://en.wikipedia.org/w/index.php?title=File:World_population_density_1994.png *License*: unknown *Contributors*: User:QuartierLatin1968

Image:Population density with key.png *Source*: http://en.wikipedia.org/w/index.php?title=File:Population_density_with_key.png *License*: unknown *Contributors*: User:Keenan Pepper, User:SG

File:Aavikko.png *Source*: http://en.wikipedia.org/w/index.php?title=File:Aavikko.png *License*: unknown *Contributors*: User Vzb83 on fi.wikipedia

Image:Monaco by night.JPG *Source*: http://en.wikipedia.org/w/index.php?title=File:Monaco_by_night.JPG *License*: unknown *Contributors*: User:Ewa Dryjanska

File:TariatLandscape.jpg *Source*: http://en.wikipedia.org/w/index.php?title=File:TariatLandscape.jpg *License*: unknown *Contributors*: User:Doron

File:EU Pop2008 1024.PNG *Source*: http://en.wikipedia.org/w/index.php?title=File:EU_Pop2008_1024.PNG *License*: unknown *Contributors*: User:Vinny Burgoo

File:Flag of Singapore.svg *Source*: http://en.wikipedia.org/w/index.php?title=File:Flag_of_Singapore.svg *License*: unknown *Contributors*: Various

File:Flag of Hong Kong.svg *Source*: http://en.wikipedia.org/w/index.php?title=File:Flag_of_Hong_Kong.svg *License*: unknown *Contributors*: Designed by

File:Flag of Bahrain.svg *Source*: http://en.wikipedia.org/w/index.php?title=File:Flag_of_Bahrain.svg *License*: unknown *Contributors*: User:SKopp, User:Zscout370

File:Flag of Bangladesh.svg *Source*: http://en.wikipedia.org/w/index.php?title=File:Flag_of_Bangladesh.svg *License*: unknown *Contributors*: User:SKopp

File:Flag of the Republic of China.svg *Source*: http://en.wikipedia.org/w/index.php?title=File:Flag_of_the_Republic_of_China.svg *License*: unknown *Contributors*: 555, Abner1069, Bestalex, Bigmorr, Bjankuloski06en, Denelson83, Ed veg, Gzdavidwong, Herbythyme, Isletakee, Kakoui, Kallerna, Kibinsky, MAXXX-309, Mattes, Mizunoryu, Neq00, Nickpo, Nightstallion, Odder, Pymouss, R.O.C, Reisio, Reuvenk, Rkt2312, Rocket000, Runningfridgesrule, Samwingkit, Sasha Krotov, Shizhao, SiBr4, Tabasco, Theo10011, Vzb83, Wrightbus, ZooFari, Zscout370, 75 anonymous edits

File:Flag of Mauritius.svg *Source*: http://en.wikipedia.org/w/index.php?title=File:Flag_of_Mauritius.svg *License*: unknown *Contributors*: User:Zscout370

File:Flag of South Korea.svg *Source*: http://en.wikipedia.org/w/index.php?title=File:Flag_of_South_Korea.svg *License*: unknown *Contributors*: Various

File:Flag of Rwanda.svg *Source*: http://en.wikipedia.org/w/index.php?title=File:Flag_of_Rwanda.svg *License*: unknown *Contributors*: .

File:Flag of the Netherlands.svg *Source*: http://en.wikipedia.org/w/index.php?title=File:Flag_of_the_Netherlands.svg *License*: unknown *Contributors*: User:Zscout370

File:Flag of Lebanon.svg *Source*: http://en.wikipedia.org/w/index.php?title=File:Flag_of_Lebanon.svg *License*: unknown *Contributors*: Traced based on the CIA World Factbook with some modification done to the colours based on information at Vexilla mundi.

File:Flag of India.svg *Source*: http://en.wikipedia.org/w/index.php?title=File:Flag_of_India.svg *License*: unknown *Contributors*: Anomie, Mifter

File:Flag of Belgium (civil).svg *Source*: http://en.wikipedia.org/w/index.php?title=File:Flag_of_Belgium_(civil).svg *License*: unknown *Contributors*: Bean49, Cathy Richards, David Descamps, Dbenbenn, Denelson83, Evanc0912, Fry1989, Gabriel trzy, Howcome, IvanOS, Mimich, Ms2ger, Nightstallion, Oreo Priest, Ricordisamoa, Rocket000, Rodejong, SiBr4, Sir Iain, ThomasPusch, Warddr, Zscout370, 6 anonymous edits

File:Flag of Japan.svg *Source*: http://en.wikipedia.org/w/index.php?title=File:Flag_of_Japan.svg *License*: unknown *Contributors*: Anomie

File:Flag of Sri Lanka.svg *Source*: http://en.wikipedia.org/w/index.php?title=File:Flag_of_Sri_Lanka.svg *License*: unknown *Contributors*: Zscout370

File:Volkstelling 1925 Census.jpg *Source*: http://en.wikipedia.org/w/index.php?title=File:Volkstelling_1925_Census.jpg *License*: unknown *Contributors*: Aiko, Classical geographer, Frank C. Müller, JPS68, Vincent Steenberg, Vluebben

File:Percentage population living on less than $1.25 per day 2009.svg *Source*: http://en.wikipedia.org/w/index.php?title=File:Percentage_population_living_on_less_than_$1.25_per_day_2009.svg *License*: unknown *Contributors*: Tony0106

File:Percent poverty world map.png *Source*: http://en.wikipedia.org/w/index.php?title=File:Percent_poverty_world_map.png *License*: unknown *Contributors*: User:Sbw01f

Image:Time zone chicago.jpg *Source*: http://en.wikipedia.org/w/index.php?title=File:Time_zone_chicago.jpg *License*: unknown *Contributors*: User:JoeSmack

File:Standard time zones of the world.png *Source*: http://en.wikipedia.org/w/index.php?title=File:Standard_time_zones_of_the_world.png *License*: unknown *Contributors*: User:TimeZonesBoy

File:UTC hue4map X world Robinson.png *Source*: http://en.wikipedia.org/w/index.php?title=File:UTC_hue4map_X_world_Robinson.png *License*: unknown *Contributors*: User:Phoenix B 1of3

Image:Tzdiff-Europe-summer.png *Source*: http://en.wikipedia.org/w/index.php?title=File:Tzdiff-Europe-summer.png *License*: unknown *Contributors*: Jmabel, Naive cynic

Image:DaylightSaving-World-Subdivisions.png *Source*: http://en.wikipedia.org/w/index.php?title=File:DaylightSaving-World-Subdivisions.png *License*: unknown *Contributors*: Paul Eggert

File:Coventry Time Zone Clock.jpg *Source*: http://en.wikipedia.org/w/index.php?title=File:Coventry_Time_Zone_Clock.jpg *License*: unknown *Contributors*: User:Cmglee

MIX
Papier aus verantwortungsvollen Quellen
Paper from responsible sources
FSC® C105338

Printed by Books on Demand GmbH, Norderstedt / Germany